Spacehounds Of IPC

by

E. E. Smith

Double 9 BOOKS

Spacehounds of IPC
by E. E. Smith

ISBN: 978-93-59328-22-5

Published by

DOUBLE 9 BOOKS
2/13-B, Ansari Road, Daryaganj
New Delhi - 110002
info@double9books.com
www.double9books.com
Tel. 011-40042856

ABOUT THE AUTHOR

Edward Elmer Smith (May 2, 1890 - August 31, 1965) was an American food engineer and science-fiction author best known for his Lensman and Skylark novels. He is sometimes referred to as the "Father of Space Opera." Edward Elmer Smith was born on May 2, 1890, in Sheboygan, Wisconsin, to Fred Jay Smith and Caroline Mills Smith, both devout Presbyterians of British origin. His mother was a teacher born in Michigan in February 1855, and his father was a sailor born in Maine to an English father in January 1855. The winter after Edward Elmer was born, they relocated to Spokane, Washington, where Mr. Smith worked as a contractor in 1900. The family relocated to Seneaquoteen, on the Pend Oreille River in Kootenai County, Idaho, in 1902. He had four siblings: Rachel M. (born September 1882), Daniel M. (born January 1884), Mary Elizabeth (born February 1886) and Walter E. (born July 1891 in Washington). Fred and Caroline Smith, together with their son Walter, were living in the Markham Precinct of Bonner County, Idaho in 1910; Fred is classified as a farmer in census data.

CONTENTS

PART I

A GOOD many of us, who are now certain beyond a doubt that space travel will forever remain in the realm of the impossible, probably would, if a rocket that were shot to the moon, for instance, did arrive, and perhaps return to give proof of its safe arrival on our satellite, accept the phenomenon in a perfectly blasé, twentieth century manner. Dr. Smith, that phenomenal writer of classic scientific fiction, seems to have become so thoroughly convinced of the advent of interplanetary travel that it is difficult for the reader to feel, after finishing "Spacehounds of IPC," that travel in the great spaces is not already an established fact. Dr. Smith, as a professional chemist, is kept fairly busy. As a writer, he is satisfied with nothing less than perfection. For that reason, a masterpiece from his pen has become almost an annual event. We know you will like "Spacehounds" even better than the "Skylark" series.

CHAPTER I
The IPV Arcturus Sets Out for Mars

ANARROW football of steel, the Interplanetary Vessel *Arcturus* stood upright in her berth in the dock like an egg in its cup. A hundred feet across and a hundred and seventy feet deep was that gigantic bowl, its walls supported by the structural steel and concrete of the dock and lined with hard-packed bumper-layers of hemp and fibre. High into the air extended the upper half of the ship of space—a sullen gray expanse of fifty-inch hardened steel armor, curving smoothly upward to a needle prow. Countless hundred of fine vertical scratches marred every inch of her surface, and here and there the stubborn metal was grooved and scored to a depth of inches—each scratch and score the record of an attempt of some wandering cosmic body to argue the right-of-way with the stupendous mass of that man-made cruiser of the void.

A burly young man made his way through the throng about the entrance, nodded unconcernedly to the gatekeeper, and joined the stream of passengers flowing through the triple doors of the double air-lock and down a corridor to the center of the vessel. However, instead of entering one of the elevators which were whisking the passengers up to their staterooms in the upper half of the enormous football, he in some way caused an opening to appear in an apparently blank steel wall and stepped through it into the control room.

"Hi, Breck!" the burly one called, as he strode up to the instrument-desk of the chief pilot and tossed his bag carelessly into a corner. "Behold your computer in the flesh! What's all this howl and fuss about poor computation?"

"Hello, Steve!" The chief pilot smiled as he shook hands cordially. "Glad to see you again—but don't try to kid the old man. I'm simple enough to believe almost anything, but some things just aren't being done. We have been yelling, and yelling hard, for trained computers ever since they started riding us about every one centimeter change in acceleration, but I know that you're no more an I-P computer than I am a Digger Indian. They don't shoot sparrows with coast-defense guns!"

"Thanks for the compliment, Breck, but I'm your computer for this trip, anyway. Newton, the good old egg, knows what you fellows are up against and is going to do something about it, if he has to lick all the rest of the directors to do it. He knew that I was loose for a couple of weeks and asked me to come along this trip to see what I could see. I'm to check the observatory data—they don't know I'm aboard—take the peaks and valleys off your acceleration curve, if possible, and report to Newton just what I find out and what I think should be done about it. How early am I?" While the newcomer was talking, he had stripped the covers from a precise scale model of the solar system and from a large and complicated calculating machine and had set to work without a wasted motion or instant—scaling off upon the model the positions of the various check-stations and setting up long and involved integrals and equations upon the calculator.

The older man studied the broad back of the younger, bent over his computations, and a tender, almost fatherly smile came over his careworn face as he replied:

"Early? You? Just like you always were—plus fifteen seconds on the deadline. The final dope is due right now." He plugged the automatic recorder and speaker into a circuit marked "Observatory," waited until a tiny light above the plug flashed green, and spoke.

"IPV *Arcturus;* Breckenridge, Chief Pilot; trip number forty-three twenty-nine. Ready for final supplementary route and flight data, Tellus to Mars."

"Meteoric swarms still too numerous for safe travel along the scheduled route," came promptly from the speaker. "You must stay further away from the plane of the ecliptic. The ether will be clear for you along route E2-P6-W41-K3-R19-S7-M14. You will hold a constant acceleration of 981.27 centimeters between initial and final check stations. Your take-off will be practically unobstructed, but you will have to use the utmost caution in landing upon Mars, because in order to avoid a weightless detour and a loss of thirty-one minutes, you must pass very close to both the Martian satellites. To do so safely you must pass the last meteorological station, M14, on schedule time plus or minus five seconds, at scheduled velocity plus or minus ten meters, with exactly the given negative acceleration of 981.27 centimeters, and exactly upon the pilot ray M14 will have set for you."

"All x." Breckenridge studied his triplex chronometer intently, then unplugged and glanced around the control room, in various parts of which half a dozen assistants were loafing at their stations.

"Control and power check-out—Hipe!" he barked. "Driving converters and projectors!"

The first assistant scanned his meters narrowly as he swung a multi-point switch in a flashing arc. "Converter efficiency 100, projector reactivity 100; on each of numbers one to forty-five inclusive. All x."

"Dirigible projectors!"

TWO more gleaming switches leaped from point to point. "Converter efficiency 100, projector reactivity 100, dirigibility 100, on each of numbers one to thirty-two, inclusive, of upper band; and numbers one to thirty-two, inclusive, of lower band. All x."

"Gyroscopes!"

"35,000. Drivers in equilibrium at ten degrees plus. All x."

"Upper lights and lookout plates!"

The second assistant was galvanized into activity, and upon a screen before him there appeared a view as though he were looking directly upward from the prow of the great vessel. The air above them was full of aircraft of all shapes and sizes, and occasionally the image of one of that flying horde flared into violet splendor upon the screen as it was caught in the mighty, roving beam of one of the twelve ultra-light projectors under test.

"Upper lights and lookout plates—all x," the second assistant reported, and other assistants came to attention as the check-out went on.

"Lower lights and lookout plates!"

"All x," was the report, after each of the twelve ultra-lights of the stern had swung around in its supporting brackets, illuminating every recess of the dark depths of the bottom well of the berth and throwing the picture upon another screen in lurid violet relief.

"Lateral and vertical detectors!"

"Laterals XP2710—all x. Verticals AJ4290—all x."

"Receptors!"

"15,270 kilofranks—all x."

"Accumulators!"

"700,000 kilofrank-hours—all x."

Having thus checked and tested every function of his department, Breckenridge plugged into "Captain," and when the green light went on:

"Chief pilot check-out—all x," he reported briefly.

"All x," acknowledged the speaker, and the chief pilot unplugged. Fifteen minutes remained, during which time one department head after another would report to the captain of the liner that everything in his charge was ready for the stupendous flight.

"All x, Steve?" Breckenridge turned to the computer. "How do you check acceleration and power with the observatory?"

"Not so good, old bean," the younger man frowned in thought. "They figure like astronomers, not navigators. They've made no allowances for anything, not even the reversal—and I figure four thousands for that and for minor detours. Then there's check station errors...."

"Check-station errors! Why, they're always right—that's what they're for!"

"Don't fool yourself—they've got troubles of their own, the same as anybody else. In fact, from a study of the charts of the last few weeks, I'm pretty sure that E2 is at least four thousand kilometers this side of where he thinks he is, that W41 is ten or twelve thousand beyond his station, and that they've both got a lateral displacement that's simply fierce. I'm going to check up, and argue with them about it as we pass. Then there's another thing—they figure to only two places, and we've got to have the third place almost solid if we expect to get a smooth curve. A hundredth of a centimeter of acceleration means a lot on a long trip when they're holding us as close as they are doing now. We'll ride this trip on 981.286 centimeters—with our scheduled mass, that means thirty six points of four seven kilofranks *plus* equilibrium power. All set to go," the computer stated, as he changed, by fractions of arc, the course-plotters of the automatic integrating goniometer.

"You're the doctor—but I'm glad it's you that'll have to explain to the observatory," and Breckenridge set his exceedingly delicate excess power potentiometer exactly upon the indicated figure. "Well, we've got a few minutes left for a chin-chin before we lift her off."

"What's all this commotion about? Dish out the low-down."

"Well, it's like this, Steve. We pilots are having one sweet time—we're being growled at on every trip. The management squawks if we're thirty seconds plus or minus at the terminals, and the passenger department squalls if we change acceleration five centimeters total en route—claims it upsets the dainty customers and loses business for the road. They're tightening up on us all the time. A couple of years ago, you remember, it didn't make any difference what we did with the acceleration as long as we checked in somewhere near zero time—we used to spin 'em dizzy when we reversed at the half-way station—but that kind of stuff doesn't go any more. We've got to hold the acceleration constant and close to normal, got to hold our schedule on zero, *plus* or *minus* ten seconds, and yet we've got to make any detours they tell us to, such as this seven-million kilometer thing they handed us just now. To make things worse, we've got to take orders at every check-station, and yet *we* get the blame for everything that happens as a consequence of obeying those orders! Of course, I know as well as you do that it's rotten technique to change acceleration at every check-station; but we've told 'em over and over that we can't do any better until they put a real computer on every ship and tell the check-stations to report meteorites and other obstructions to us and then to let us alone. So you'd better recommend us some computers!"

"You're getting rotten computation, that's a sure thing, and I don't blame you pilots for yelling, but I don't believe that you've got the right answer. I can't help but think that the astronomers are lying down on the job. They are so sure that you pilots are to blame that it hasn't occurred to them to check up on themselves very carefully. However, we'll know pretty quick, and then we'll take steps."

"I hope so—but say, Steve, I'm worried about using that much plus equilibrium power. Remember, we've got to hit M14 in absolutely good shape, or plenty heads will drop."

"I'll say they will. I know just how the passengers will howl if we hold them weightless for half an hour, waiting for those two moons to get out of the way, and I know just what the manager will do if we check in minus thirty-one minutes. Wow! He'll swell up and bust, sure. But don't worry,

Breck—if we don't check in all right, anybody can have my head that wants it, and I'm taking full responsibility, you know."

"You're welcome to it." Breckenridge shrugged and turned the conversation into a lighter vein. "Speaking of weightlessness, it's funny how many weight-fiends there are in the world, isn't it? You'd think the passengers would enjoy a little weightlessness occasionally—especially the fat ones—but they don't. But say, while I think of it, how come you were here and loose to make this check-up? I thought you were out with the other two of the Big Three, solving all the mysteries of the Universe?"

"Had to stay in this last trip—been doing some work on the ether, force-field theory, and other advanced stuff that I had to go to Mars and Venus to get. Just got back last week. As for solving mysteries, laugh while you can, old hyena. You and a lot of other dim bulbs think that Roeser's Rays are the last word—that there's nothing left to discover—are going to get jarred loose from your hinges one of these days. When I came in nine months ago they were hot on the trail of something big, and I'll bet they bring it in...."

Out upon the dock an insistent siren blared a crescendo and diminuendo blast of sound, and two minutes remained. In every stateroom and in every lounge and saloon speakers sounded a warning:

"For a short time, while we are pulling clear of the gravitational field of the Earth, walking will be somewhat difficult, as everything on board will apparently increase in weight by about one-fifth of its present amount. Please remain seated, or move about with caution. In about an hour weight will gradually return to normal. We start in one minute."

"Hipe!" barked the chief pilot as a flaring purple light sprang into being upon his board, and the assistants came to attention at their stations. "Seconds! Four! Three! Two! One! LIFT!" He touched a button and a set of plunger switches drove home, releasing into the forty-five enormous driving projectors the equilibrium power—the fifteen-thousand-and-odd kilofranks of energy that exactly counterbalanced the pull of gravity upon the mass of the cruiser. Simultaneously there was added from the potentiometer, already set to the exact figure given by the computer, the *plus*-equilibrium power—which would not be changed throughout the journey if the ideal acceleration curve were to be registered upon the recorders—and the immense mass of the cruiser of the void wafted vertically upward at a low and constant velocity. The bellowing, shrieking siren had cleared the air magically of the swarm of aircraft in her path, and quietly, calmly, majestically, the *Arcturus* floated upward.

BRECKENRIDGE, sixty seconds after the initial lift, actuated the system of magnetic relays which would gradually cut in the precisely measured "starting power," which it would be necessary to employ for sixty-nine minutes—for, without the acceleration given by this additional power, they would lose many precious hours of time in covering merely the few thousands of miles during which Earth's attraction would operate powerfully against their progress.

Faster and faster the great cruiser shot upward as more and more of the starting power was released, and heavier and heavier the passengers felt themselves become. Soon the full calculated power was on and the acceleration became constant. Weight no longer increased, but remained constant at a value of plus twenty three and six-tenths percent. For a few moments there had been uneasy stomachs among the passengers—perhaps a few of the first-trippers had been made ill—but it was not much worse than riding in a high-speed elevator, particularly since there was no change from positive to negative acceleration such as is experienced in express elevators.

The computer, his calculations complete, watched the pilot with interest, for, accustomed as he was to traversing the depths of space, there was a never-failing thrill to his scientific mind in the delicacy and precision of the work which Breckenridge was doing—work which could be done only by a man who had had long training in the profession and who was possessed of instantaneous nervous reaction and of the highest degree of manual dexterity and control. Under his right and left hands were the double-series potentiometers actuating the variable-speed drives of the flight-angle directors in the hour and declination ranges; before his eyes was the finely marked micrometer screen upon which the guiding goniometer threw its needle-point of light; powerful optical systems of prisms and lenses revealed to his sight the director-angles, down to fractional seconds of arc. It was the task of the chief pilot to hold the screened image of the cross-hairs of the two directors in such position relative to the ever-moving point of light as to hold the mighty vessel precisely upon its course, in spite of the complex system of forces acting upon it.

For almost an hour Breckenridge sat motionless, his eyes flashing from micrometer screen to signal panel, his sensitive fingers moving the potentiometers through minute arcs because of what he saw upon the screen and in instantaneous response to the flashing, multi-colored lights and tinkling signals of his board. Finally, far from earth, the moon's attraction and other perturbing forces comparatively slight, the signals no longer sounded and the point of light ceased its irregular motion, becoming almost stationary. The chief pilot brought both cross-hairs directly upon the brilliant point, which for some time they had been approaching more and more nearly, adjusted the photo-cells and amplifiers which would hold them immovably upon it, and at the calculated second of time, cut out the starting power by means of another set of automatically timed relays. When only the regular driving power was left, and the acceleration had been checked and found to be exactly the designated value of 981.286 centimeters, he stood up and heaved a profound sigh of relief.

"Well, Steve, that's over with—we're on our way. I'm always glad when this part of it is done."

"It's a ticklish job, no fooling—even for an expert," the mathematician agreed. "No wonder the astronomers think you birds are the ones who are gumming up their dope. Well, it's about time to plug in on E2. Here's where the fireworks start!" He closed the connections which transferred the central portion of the upper lookout screen to a small micrometer screen at Breckenridge's desk and plugged it into the first check-station. Instantly a point of red light, surrounded by a vivid orange circle, appeared upon the screen, low down and to the left of center, and the timing galvanometer showed a wide positive deflection.

"Hashed again!" growled Breckenridge. "I must be losing my grip, I guess. I put everything I had on that sight, and missed it ten divisions. I think I'll turn in my badge—I've cocked our perfect curve already, before we got to the first check-station!" His hands moved toward the controls, to correct their course and acceleration.

"As you were—hold everything! Lay off those controls!" snapped the computer. "There's something screwy, just as I thought—and it isn't you, either. I'm no pilot, of course, but I do know good compensation when I see it, and if you weren't compensating that point I never saw it done. Besides, with your skill and my figures I know darn well that we aren't off more than a tenth of one division. He's cuckoo! Don't call him—let him start it, and refer him to me."

"All x—I'll be only too glad to pass the buck. But I still think, Steve, that you're playing with dynamite. Who ever heard of an astronomer being wrong?"

"You'd be surprised," grinned the physicist, "Since this fuss has just started, nobody has tried to find out whether they were wrong or not...."

"IPV *Arcturus*, attention!" came from the speaker curtly.

"IPV *Arcturus*, Breckenridge," from the chief pilot.

"You have been on my ray almost a minute. Why are you not correcting course and acceleration?"

"Doctor Stevens is computing us and has full control of course and acceleration," replied Breckenridge. "He will answer you."

"I am changing neither course nor acceleration because you are not in position," declared Stevens, crisply, "Please give me your present supposed location, and your latest precision goniometer bearings on the sun, the moon, Mars, Venus, and your Tellurian reference limb, with exact time of observations, gyroscope zero-planes, and goniometer factors!"

"Correct at once or I shall report you to the Observatory," E2 answered loftily, paying no attention to the demand for proof of position.

"Be sure you do that, guy—and while you're at it report that your station hasn't taken a precision bearing in a month. Report that you've been muddling along on radio loop bearings, and that you don't know where you are, within seven thousand kilometers. And speaking of reporting—I know already that a lot of you astronomical guessers have only the faintest possible idea of where you really are, *plus*, *minus*, or lateral; and if you don't get yourselves straightened out before we get to W41, I'm going to make a report on my own account that will jar some of you birds loose from your upper teeth!" He unplugged with a vicious jerk, and turned to the pilot with a grin.

"Guess that'll hold him for a while, won't it?"

"He'll report us, sure," remonstrated Breckenridge. The older man was plainly ill at ease at this open defiance of the supposedly infallible check-stations.

"Not that baby," returned the computer confidently. "I'll bet you a small farm against a plugged nickel that right now he's working his goniometer so hard that it's pivots are getting hot. He'll sneak back into position as soon as he can calculate his results, and pretend he's always been there."

"The others will be all right, then, probably, by the time we get to them?"

"Gosh, no—you're unusually dumb today, Breck. He won't tell anybody anything—he doesn't want to be the only goat, does he?"

"Oh, I see. How could you dope this out, with only the recorder charts?"

"Because I know the kind of stuff you pilots are—and those humps are altogether too big to be accounted for by anything I know about you. Another thing—the next station, P6, I think is keeping himself all x. If so, when you corrected for E2, which was wrong, it'd throw you all off on P6, which was right, and so on—a bad hump at almost every check-station. See?"

TRUE to prediction, the pilot ray of P6 came in almost upon the exact center of the micrometer screen, and Breckenridge smiled in relief as he began really to enjoy the trip.

"How do we check on chronometers?" asked P6 when Stevens had been introduced. "By my time you seem to be about two and a half seconds *plus*?"

"All x—two points four seconds plus—we're riding on 981.286 centimeters, to allow for the reversal and for minor detours. Bye."

"All this may have been coincidence, Breck, but we'll find out pretty quick now," the computer remarked when the flying vessel was nearing the third check-station. "Unless I'm all out of control we'll check in almost fourteen seconds minus on W41, and we may not even find him on the center block of the screen."

When he plugged in W41 was on the block, but was in the extreme upper right corner. They checked in thirteen and eight-tenths seconds minus on the station, and a fiery dialogue ensued when the computer questioned the accuracy of the location of the station and refused point-blank to correct his course.

"Well, Breck, old onion, that tears it," Stevens declared as he unplugged. "No use going any further on these bum reference points. I'm going to report to Newton—he'll rock the Observatory on its foundations!" He plugged into the telegraph room. "Have you got a free high-power wave?... Please put me on Newton, in the main office."

Moving lights flashed and flickered for an instant upon the communicator screen, settling down into a white glow which soon resolved itself into the likeness of a keen-eyed, gray-haired man, seated at his desk in the remote office of the Interplanetary Corporation. Newton smiled as he recognized the likeness of Stevens upon his own screen, and greeted him cordially.

"Have you started your investigation, Doctor Stevens?"

"Started it? I've finished it!" and Stevens tersely reported what he had learned, concluding: "So you see, you don't need special computers on these ships any more than a hen needs teeth. You've got all the computers you need, in the observatories—all you've got to do is make them work at their trade."

"The piloting was all x, then?"

"Absolutely—our curve so far is exactly flat ever since we cut off the starting power. Of course, all the pilots can't be as good as Breckenridge, but give them good computation and good check points and you shouldn't get any humps higher than about half a centimeter."

"They'll get both, from now on," the director assured him. "Thanks. If your work for the trip is done, you might show my little girl, Nadia, around the *Arcturus*. She's never been out before, and will be interested. Would you mind?"

"Glad to, Mr. Newton—I'll be a regular uncle to her."

"Thanks again, Operator, I'll speak to Captain King, please."

"Pipe down that guff, you unlicked cub, or I'll crown you with a proof-bar!" the chief pilot growled, as soon as Stevens had unplugged.

"You and who else?" retorted the computer, cheerfully. "Pipe down yourself, guy—if you weren't so darn dumb and didn't have such a complex, you'd know that you're the crack pilot of the outfit and wouldn't care who else knew it." Stevens carefully covered and put away the calculating machine and other apparatus he had been using and turned again to the pilot.

"I didn't know Newton had any kids, especially little ones, or I'd have got acquainted with them long ago. Of course I don't know him very well, since I never was around the office much, but the old tiger goes over big with me."

"Hm—m. Think you'll enjoy playing nursemaid all the rest of the trip?" Breckenridge asked caustically, but with an enigmatic smile.

"Think so? I *know* so!" replied Stevens, positively. "I always did like kids, and they always did like me—we fall for each other like ten thousand bricks falling down a well. Why, a kid—*any* kid—and I team up just like grace and poise.... What's gnawing on you anyway, to make you turn Cheshire cat all of a sudden? By the looks of that grin I'd say you had swallowed a canary of mine some way or other; but darned if I know that I've lost any," and he stared at his friend suspiciously.

"To borrow your own phrase, Steve, 'You'd be surprised,'" and Breckenridge, though making no effort to conceal his amusement, would say no more.

In a few minutes the door opened, and through it there stepped a grizzled four-striper. Almost hidden behind his massive form there was a girl, who ran up to Breckenridge and seized both his hands, her eyes sparkling.

"Hi, Breckie, you old darling! I knew that if we both kept after him long enough Dad would let me ride with you sometime. Isn't this *gorgeous*?"

Stevens was glad indeed that the girl's enthusiastic greeting of the pilot was giving him time to recover from his shock, for Director Newton's "little girl, Nadia" was not precisely what he had led himself to expect. Little she might be, particularly when compared with the giant frame of Captain King, or with Steve's own five-feet-eleven of stature and the hundred and ninety pounds of rawhide and whalebone that was his body, but child she certainly was not. Her thick, fair hair, cut in the square bob that was the mode of the moment, indicated that Nature had intended her to be a creamy blonde, but as she turned to be introduced to him, Stevens received another surprise—for she was one of those rare, but exceedingly attractive beings, a natural blonde with brown eyes and black eyebrows. Sun and wind had tanned her satin skin to a smooth and even shade of brown, and every movement of her lithe and supple body bespoke to the discerning mind a rigidly-trained physique.

"Doctor Stevens, you haven't met Miss Newton, I hear," the captain introduced them informally. "All the officers who are not actually tied down at their posts are anxious to do the honors of the vessel, but as I have received direct orders from the owners, I am turning her over to you—you are to show her around."

"Thanks, Captain, I won't mutiny a bit against such an order. I'm mighty glad to know you, Miss Newton."

"I've heard a lot about you, Doctor. Dad and Breckie here are always talking about the Big Three—what you have done and what you are going to do. I want to meet Doctor Brandon and Doctor Westfall, too," and her hand met his in a firm and friendly clasp. She turned to the captain, and Stevens, noticing that the pilot, with a quizzical expression, was about to say something, silenced him with a fierce aside.

"Clam it, ape, or I'll climb up you like a squirrel!" he hissed, and the grinning Breckenridge nodded assent to this demand for silence concerning children and nursemaids.

"Since you've never been out, Miss Newton, you'll want to see the whole works," Stevens addressed the girl. "Where do you want to begin? Shall we start at the top and work down?"

"All right with me," she agreed, and fell into step beside him. She was dressed in dove-gray from head to foot—toque, blouse, breeches, heavy stockings, and shoes were of the one shade of smooth, lustrous silk; and as they strolled together down the passage-way, the effortless ease and perfect poise of her carriage called aloud to every hard-schooled fibre of his own highly-trained being.

"We're a lot alike you and I—do you know it?" he asked, abruptly and unconventionally.

"Yes, I've felt it, too," she replied frankly, and studied him without affectation. "It has just come to me what it is. We're both in fine condition and in hard training. You're an athlete of some kind, and I'm sure you're a star—I ought to recognize you, but I'm ashamed to say I don't. What do you do?"

"Swim."

"Oh, of course—Stevens, the great Olympic high and fancy diver! I would *never* have connected our own Doctor Stevens, the eminent mathematical physicist, with the King of the Springboard. Say, ever since I quit being afraid of the water I've had a yen to do that two-and-a-half twist of yours, but I never met anybody who knew it well enough to teach it to me, and I've almost broken my back forty times trying to learn it alone!"

"I've got you, now, too—American and British Womens' golf champion. Shake!" and the two shook hands vigorously, in mutual congratulation. "Tell you what—I'll give you some pointers on diving, and you can show me how to make a golf ball behave. Next to Norman Brandon, I've got the most vicious hook in captivity—and Norm can't help himself. He's left-handed, you know, and, being a southpaw, he's naturally wild. He slices all his woods and hooks all his irons. I'm consistent, anyway—I hook everything, even my putts."

"It's a bargain! What do you shoot?"

"Pretty dubby. Usually in the middle eighties—none of us play much, being out in space most of the time, you know—sometimes, when my hook is going particularly well, I go up into the nineties."

"We'll lick that hook," she promised, as they entered an elevator and were borne upward, toward the prow of the great interplanetary cruiser.

CHAPTER II
But Does Not Arrive

"**A**LL out—we climb the rest of the way on foot," Stevens told his companion, as the elevator stopped at the uppermost passenger floor. They walked across the small circular hall and the guard on duty came to attention and saluted as they approached him.

"I have orders to pass you and Miss Newton, sir. Do you know all the combinations?"

"I know this good old tub better than the men that built her—I helped calculate her," Stevens replied, as he stepped up to an apparently blank wall of steel and deftly manipulated an almost invisible dial set flush with its surface. "This is to keep the passengers where they belong," he explained, as a section of the wall swung backward in a short arc and slid smoothly aside. "We will now proceed to see what makes it tick."

Ladder after ladder of steel they climbed, and bulkhead after bulkhead opened at Stevens's knowing touch. At each floor the mathematician explained to the girl the operation of the machinery there automatically at work—devices for heating and cooling, devices for circulating, maintaining, and purifying the air and the water—in short, all the complex mechanism necessary for the comfort and convenience of the human cargo of the liner.

Soon they entered the conical top compartment, a room scarcely fifteen feet in diameter, tapering sharply upward to a hollow point some twenty feet above them. The true shape of the room, however, was not immediately apparent, because of the enormous latticed beams and girders which braced the walls in every direction. The air glowed with the violet light of the twelve great ultra-light projectors, like searchlights with three-foot lenses, which lined the wall. The floor beneath their feet was not a level steel platform, but seemed to be composed of many lenticular sections of dull blue alloy.

"We are standing upon the upper lookout lenses, aren't we?" asked the girl. "Is that perfectly all right?"

"Sure. They're so hard that nothing can scratch them, and of course Roeser's Rays go right through our bodies, or any ordinary substance, like a bullet through a hole in a Swiss cheese. Even those lenses wouldn't deflect them if they weren't solid fields of force."

As he spoke, one of the ultra-lights flashed around in a short, quick arc, and the girl saw that instead of the fierce glare she had expected, it emitted only a soft violet light. Nevertheless she dodged involuntarily and Stevens touched her arm reassuringly.

"All x, Miss Newton—they're as harmless as mice. They hardly ever have to swing past the vertical, and even if one shines right through you you can look it right in the eye as long as you want to—it can't hurt you a bit."

"No ultra-violet at all?"

"None whatever. Just a color—one of the many remaining crudities of our ultra-light vision. A lot of good men are studying this thing of direct vision, though, and it won't be long before we have a system that will really work."

"I think it's all perfectly wonderful!" she breathed. "Just think of traveling in comfort through empty space, and of actually seeing through seamless steel walls, without even a sign of a window! How can such things be possible?"

"I'll have to go pretty well back," he warned, "and any adequate explanation is bound to be fairly deep wading in spots. How technical can you stand it?"

"I can go down with you middling deep—I took a lot of general science, and physics through advanced mechanics. Of course, I didn't get into any such highly specialized stuff as sub-electronics or Roeser's Rays, but if you start drowning me, I'll yell."

"That's fine—you can get the idea all x, with that to go on. Let's sit down here on this girder. Roeser didn't do it all, by any means, even though he got credit for it—he merely helped the Martians do it. The whole thing started, of course, when Goddard shot his first rocket to the moon, and was intensified when Roeser so perfected his short waves that signals were exchanged with Mars—signals that neither side could make any sense out of. Goddard's pupils and followers made bigger and better rockets, and finally got one that could land safely upon Mars. Roeser, who was a mighty keen bird, was one of the first voyagers, and he didn't come back—he stayed there, living in a space-suit for three or four years, and got a brand-new education. Martian science always was hot, you know, but they were impractical. They were desperately hard up for water and air, and while they had a lot of wonderful ideas and theories, they couldn't overcome the practical technical difficulties in the way of making their ideas work. Now putting other peoples' ideas to work was Roeser's long suit—don't think that I'm belittling Roeser at all, either, for he was a brave and far-sighted man, was no mean scientist, and was certainly one of the best organizers and synchronizers the world has ever known—and since Martian and Tellurian science complemented each other, so that one filled in the gaps of the other, it wasn't long until fleets of space-freighters were bringing in air and water from Venus, which had more of both than she needed or wanted.

"Having done all he could for the Martians and having learned most of the stuff he wanted to know, Roeser came back to Tellus and organized Interplanetary, with scientists and engineers on all three planets, and set to work to improve the whole system, for the vessels they used then were dangerous—regular mankillers, in fact. At about this same time Roeser and the Interplanetary Corporation had a big part in the unification of the world into one nation, so that wars could no longer interfere with progress."

"WITH this introduction I can get down to fundamentals. Molecules are particles of the first order, and vibrations of the first order include sound, light, heat, electricity, radio, and so on. Second order, atoms—extremely short vibrations, such as hard X-rays. Third order, electrons and protons, with their accompanying Millikan, or cosmic, rays. Fourth order, sub-electrons and sub-protons. These, in the material aspect, are supposed to be the particles of the fourth order, and in the energy aspect they are known as Roeser's Rays. That is, these fourth-order rays and particles seem to partake of the nature of both energy and matter. Following me?"

"Right behind you," she assured him. She had been listening intently, her wide-spaced brown eyes fastened upon his face.

"Since these Roeser's Rays, or particles or rays of the fourth order, seem to be both matter and energy, and since the rays can be converted into what is supposed to be the particles, they have been thought to be the things from which both electrons and protons were built. Therefore, everybody except Norman Brandon has supposed them the ultimate units of creation, so that it would be useless to try to go any further...."

"Why, we were taught that they *are* the ultimate units!" she protested.

"I know you were—but we really don't know anything, except what we have learned empirically, even about our driving forces. What is called the fourth-order particle is absolutely unknown, since nobody has been able to detect it, to say nothing of determining its velocity or other properties. It has been assumed to have the velocity of light only because that hypothesis does not conflict with observational data. I'm going to give you the generally accepted idea, since we have nothing definite to offer in its place, but I warn you that that idea is very probably wrong. There's a lot of deep stuff down there hasn't been dug up yet. In fact, Brandon thinks that the product of conversion isn't what we think it is, at all—that the actual fundamental unit and the primary mechanism of the transformation lie somewhere below the fourth order, and possibly even below the level of the ether—but we haven't been able to find a point of attack yet that will let us get in anywhere. However, I'm getting 'way ahead of our subject. To get back to it, energy can be converted into something that acts like matter through Roeser's Rays, and that is the empirical fact underlying the drive of our space-ships, as well as that of almost all other vehicles on all three planets. Power is generated by the great waterfalls of Tellus and Venus—water's mighty scarce on Mars, of course, so most of our plants there use fuel—and is transmitted on light beams, by means of powerful fields of force to the receptors, wherever they may be. The individual transmitting fields and receptors are really simply matched-frequency units, each matching the electrical characteristics of some particular and unique beam of force. This beam is composed of Roeser's Rays, in their energy aspect. It took a long time to work out this tight-beam transmission of power, but it was fairly simple after they got it."

He took out a voluminous notebook, at the sight of which Nadia smiled.

"A computer might forget to dress, but you'd never catch one without a full magazine pencil and a lot of blank paper," he grinned in reply and went on, writing as he talked.

"For any given frequency, *f*, and phase angle, *theta*, you integrate, between limits zero and *pi* divided by two, sine theta d...."

"Hold it—I'm sinking!" Nadia exclaimed. "I don't integrate at all unless it is absolutely necessary. As long as you stick to general science, I'm right on your heels, but please lay off of integrations and all that—most especially stay away from those terrible electrical integrations. I always did think that they were the most poisonous kind known. I want only a general idea—that's all that I can understand, anyway."

"Sure, I forgot—guess I was getting in deeper than is necessary, especially since this whole thing of beam transmission is pretty crude yet and is bound to change a lot before long. There is so much loss that when we get more than a few hundred million kilometers away from a power-plant we lose reception entirely. But to get going again, the receptors receive the beam and from them the power is sent to the accumulators, where it is stored. These accumulators are an outgrowth of the storage battery. The theory of the accumulator is...."

"Lay off the theory, please!" the listener interrupted. "I understand perfectly without it. Energy is stored in the accumulators—you put it in and take it out. That's all that is necessary."

"I'D like to give you some of the theory—but, after all, it wouldn't add much to your understanding of the working of things, and it might mix you up, as some of it is pretty deep stuff. Then, too, it would take a lot of time, and the rest of your friends would squawk if I kept you here indefinitely. From the accumulators, then, the power is fed to the converters, each of which is backed by a projector. The converters simply change the aspect of the rays, from the energy aspect to the material aspect. As soon as this is done, the highly-charged particles—or whatever they are—thus formed are repelled by the terrific stationary force maintained in the projector backing the converter. Each particle departs with a velocity supposed to be that of light, and the recoil upon the projector drives the vessel, or car, or whatever it is attached to. Still with me?"

"Struggling a little, but my nose is still above the surface. These particles, being so infinitesimally small that they cannot even be detected, go right through any substance without any effect—they are not even harmful."

"Exactly. Now we are in position to go ahead with the lights, detectors, and so on. The energy aspect of the rays you can best understand as simply a vibration in the ether—an extremely high frequency one. While not rigidly scientific, that is close enough for you and me. Nobody knows what the stuff really is, and it cannot be explained or demonstrated by any model or concept in three-dimensional space. Its physical-mathematical interpretation, the only way in which it can be grasped at all, requires sixteen coordinates in four dimensions, and I don't suppose you'd care to go into that."

"I'll say I wouldn't!" she exclaimed, feelingly.

"Well, anyway, by the use of suitable fields of force it can be used as a carrier wave. Most of this stuff of the fields of force—how to carry the modulation up and down through all the frequency changes necessary—was figured out by the Martians ages ago. Used as a pure carrier wave, with a sender and a receiver at each end, it isn't so bad—that's why our communicator and radio systems work as well as they do. They are pretty good, really, but the ultra-light vision system is something else again. Sending the heterodyned wave through steel is easy, but breaking it up, so as to view an object and return the impulses, was an awful job and one that isn't half done yet. We see things, after a fashion and at a distance of a few kilometers, by sending an almost parallel wave from a twin-projector to disintegrate and double back the viewing wave. That's the way the lookout plates and lenses work, all over the ship—from the master-screens in the control room to the plates of the staterooms and lifeboats and the viewing-areas of the promenades. But the whole system is a rotten makeshift, and...."

"Just a minute!" exclaimed the girl. "I and everybody else have been thinking that everything is absolutely perfect; and yet every single thing you have talked about, you have ended up by describing as 'unknown,' 'rudimentary,' 'temporary,' or a 'makeshift.' You speak as though the entire system were a poor thing that will have to do until something better has been found, and that nobody knows anything about anything! How do you get that way?"

"By working with Brandon and Westfall. Those birds have got real brains and they're on the track of something that will, in all probability, be as far ahead of Roeser's Rays as our present system is ahead of the science of the seventeenth century."

"Really?" she looked at him in astonishment. "Tell me about it."

"Can't be done," he refused. "I don't know much about it—even they didn't know any too much about some of it when I had to come in. And what little I do know I can't tell, because it isn't mine."

"But you're working with them, aren't you?"

"Yes, in the sense that a small boy helps his father build a house. They're the brains—I simply do some figuring that they don't want to waste time doing."

Nadia, having no belief whatever in his modest disclaimer, but in secret greatly pleased by his attitude, replied:

"Of course you couldn't say anything about an unfinished project—I shouldn't have asked. Where do we go from here?"

"Down the lining of the hull, outside the passengers' quarters to the upper dirigible projectors," and he led the way down a series of steep steel stairways, through bulkheads and partitions of steel. "One thing I forgot to tell you about—the detectors. They're worked on the same principle as the lights, and are just about as efficient. Instead, of light, though, they send out cones of electromagnetic waves, which set up induced currents in any conductor encountered beyond our own shell. Since all dangerous meteorites have been shown to contain conducting material, that is enough to locate them, for radio finders automatically determine the direction, distance, and magnitude of the disturbance, and swing a light on it. That was what happened when that light swung toward us, back there in the prow."

"Are there any of those life-boats, that I've heard discussed so much lately, near here?" asked the girl.

"Lots of 'em—here's one right here," and at the next landing he opened a vacuum-insulated steel door, snapped on a light, and waved his hand. "You can't see much of it from here, but it's a complete space-ship in itself, capable of maintaining a dozen or fifteen persons during a two-weeks' cruise in space."

"Why isn't it a good idea to retain them? Accidents are still possible, are they not?"

"Of course, and there is no question of doing away with them entirely. Modern ships, however, have only enough of them to take care of the largest number of persons ever to be carried by the vessel."

"Has the *Arcturus* more than she needs?"

"I'll say she has, and more of everything else, except room for pay-load."

"I've heard them talking about junking her. I think it's a shame."

"So do I, in a way—you see, I helped design her and her sister-ship, the *Sirius*, which Brandon and Westfall are using as a floating laboratory. But times change, and the inefficient must go. She's a good old tub, but she was built when everybody was afraid of space, and we had to put every safety factor into her that we could think of. As a result, she is four times as heavy as she should be, and that takes a lot of extra power. Her skin is too thick. She has too many batteries of accumulators, too many life-boats, too many bulkheads and air-breaks, too many and too much of everything. She is so built that if she should break up out in space, nobody would die if they lived through the shock—there are so many bulkheads, air-breaks, and life-boats that no matter how many pieces she broke up into, the survivors would find themselves in something able to navigate. That excessive construction is no longer necessary. Modern ships carry ten times the pay-load on one-quarter of the power that this old battle-wagon uses. Even though she's only four years old, she's a relic of the days when we used to slam through on the ecliptic route, right through all the meteoric stuff that is always there—trusting to heavy armor to ward off anything too small for the observers and detectors to locate. Now, with the observatories and check-stations out in space, fairly light armor is sufficient, as we route ourselves well away from the ecliptic and so miss all the heavy stuff. So, badly as I hate to see her go there, the old tub is bound for the junk-yard."

AFEW more flights of stairs brought them to the upper band of dirigible projectors, which encircled the hull outside the passengers' quarters, some sixty feet below the prow. They were heavy, search-light-like affairs mounted upon massive universal bearings, free to turn in any direction, and each having its converter nestling inside its prodigious field of force.

Stevens explained that these projectors were used in turning the vessel and in dodging meteorites when necessary, and they went on through another almost invisible door into a hall and took an elevator down to the main corridor.

"Well, you've seen it, Miss Newton," Stevens said regretfully, as he led her toward the captain's office. "The lower half is full of heavy stuff—accumulators, machinery, driving projectors, and such junk, so that the center of gravity is below the center of action of the driving projectors. That makes stable flight possible. It's all more or less like what we've just seen, and I don't suppose you want to miss the dance—anyway, a lot of people want to dance with you."

"Wouldn't you just as soon show me through the lower half as dance?"

"Rather, lots!"

"So would I. I can dance any time, and I want to see everything. Let's go!"

Down they went, past battery after battery of accumulators; climbing over and around the ever-increasing number of huge steel girders and bracers; through mazes of heavily insulated wiring and conduits; past mass after mass of automatic machinery which Stevens explained to his eager listener. They inspected one of the great driving projectors, which, built rigidly parallel to the axis of the ship and held immovably in place by enormous trusses of steel, revealed neither to the eye nor to the ear any sign of the terrific force it was exerting. Still lower they went, until the girl had been shown everything, even down to the bottom ultra-lights and stern braces.

"Tired?" Stevens asked, as the inspection was completed.

"Not very. It's been quite a climb, but I've had a wonderful time."

"So have I," he declared, positively. "I know what—we'll crawl up into one of these stern lifeboats and make us a cup of coffee before we climb back. With me?"

"'Way ahead of you!" Nadia accepted the invitation enthusiastically, and they made their way to the nearest of the miniature space-cruisers. Here, although no emergency had been encountered in all the four years of the vessel's life, they found everything in readiness, and the two soon had prepared and eaten a hearty luncheon.

"Well, I can't think of any more excuses for monopolizing you, Miss Newton, so I suppose I'll have to take you back. Believe me, I've enjoyed this more than you can realize—I've...."

He broke off and listened, every nerve taut. "What was that?" he exclaimed.

"What was what? I didn't hear anything?"

"Something screwy somewhere! I felt a vibration, and anything that'd make this mountain of steel even quiver must have given us one gosh-awful nudge. There's another!"

The girl, painfully tense, felt only a barely perceptible tremor, but the computer, knowing far better than she the inconceivable strength and mass of that enormous structure of solidly braced hardened steel, sprang into action. Leaping to the small dirigible look-out plate, he turned on the power and swung it upward.

"GREAT suffering snakes!" he ejaculated, then stood mute, for the plate revealed a terrible sight. The entire nose of the gigantic craft had been sheared off in two immense slices as though clipped off by a gigantic sword, and even as they stared, fascinated, at the sight, the severed slices were drifting slowly away. Swinging the view along the plane of cleavage, Stevens made out a relatively tiny ball of metal, only fifty feet or so in diameter, at a distance of perhaps a mile. From this ball there shot a blinding plane of light, and the *Arcturus* fell apart at the midsection, the lower half separating clean from the upper portion, which held the passengers. Leaving the upper half intact, the attacker began slicing the lower, driving half into thin, disk-shaped sections. As that incandescent plane of destruction made its first flashing cut through the body of the *Arcturus*, accompanied by an additional pyrotechnic display of severed and short-circuited high-tension leads, Stevens and Nadia suddenly found themselves floating weightless in the air of the room. Still gripping the controls of the look-out plate, Stevens caught the white-faced girl with one hand, drew her down beside him, and held her motionless while his keen mind flashed over all the possibilities of the situation and planned his course of action.

"They're apparently slicing us pretty evenly, and by the looks of things, one cut is coming right about here," he explained rapidly, as he found a flashlight and drew his companion through the door and along a narrow passage. Soon he opened another door and led her into a tiny compartment so low that they could not stand upright—a mere cubicle of steel. Carefully closing the door, he fingered dials upon each of the walls of the cell, then folded himself up into a comfortable position, instructed Nadia to do the same, and snapped off the light.

"Please leave it on," the shaken girl asked. "It's so ghastly!"

"We'd better save it, Nadia," he advised, pressing her arm reassuringly, "it's the only light we've got, and we may need it worse later on—its life is limited, you know."

"Later on? Do you think we'll need anything—later on?"

"Sure! Of course they may get us, Nadia, but this little tertiary air-break is a mighty small target for them to hit. And if they miss us, as I think they will, there's a larger room opening off each wall of this one—at least one of which will certainly be left intact. From any one of those rooms we can reach a life-boat. Of course, it's a little too much to expect that any one of the life-boats will be left whole, but they're bulkheaded, too, you know, so that we can be sure of finding something able to navigate—providing we can make our get-away. Believe me, ace, I'm sure glad we're aboard the old *Arcturus* right now, with all her safety-devices, instead of on one of the modern liners. We'd be sunk right."

"I felt sunk enough for a minute—I'm feeling better now, though, since you are taking it so calmly."

"Sure—why not? A man's not dead until his heart stops beating, you know—our turn'll come next, when they let up a little."

"But suppose they change the width of their slices, and hit this cubby, small as it is?"

"It'd be just too bad," he shrugged. "In that case, we'd never know what hit us, so it's no good worrying about it. But say, we might do something at that, if they didn't hit us square. I can move fairly fast, and might be able to get a door open before the loss of pressure seals it. We'll light the flash ... here, you hold it, so that I can have both hands free. Put both arms around me, just under the arms, and stick to me like a porous plaster, because if I have to move at all, I'll have to jump like chain lightning. Shine the beam right over there, so it'll reflect and light up all the dials at once. There ... hold on tight! Here they come!"

As he spoke, a jarring shudder shook one side of their hiding-place, then, a moment later, the phenomenon was repeated, but with much less force, upon the other side. Stevens sighed with relief, took the light, and extinguished it.

"Missed us clean!" he exulted. "Now, if they don't find us, we're all set."

"How can they possibly find us? I seem to be always worried about the wrong things, but I should think that their finding us would be the least of our troubles."

"Don't judge their vision system by ours—they've got everything, apparently. However, their apparatus may not be delicate enough to spot us in a space this small when their projectors flash through it, as they probably will. Then, too, there's a couple of other big items in our favor—nobody else is in the entire lower half, since all this machinery down here is either automatic or else controlled from up above, so they won't be expecting to see anybody when they get down this far; and we aren't at all conspicuous. We're both dressed in gray—your clothes in particular are almost exactly the color of this armor-plate—so altogether we stand a good chance of being missed."

"What shall we do now?"

"Nothing whatever—wish we could sleep for a couple of hours, but of course there's no hope of that. Stretch out here, like that—you can't rest folded up like an accordion—and I'll lie down diagonally across the room. There's just room for me that way. That's one advantage of weightlessness—you can lie down standing on your head, and go to sleep and like it. But I forgot—you've never been weightless before, have you? Does it make you sick?"

"Not so much, now, except that I feel awfully weird inside. I was horribly dizzy and nauseated at first, but it's going away."

"THAT'S good—it makes lots of people pretty sick. In fact, some folks get awfully sick and can't seem to get used to it at all. It's the canals in the inner ear that do most of it, you know. However, if you're as well as that already, you'll be a regular spacehound in half an hour. I've been weightless for weeks at a stretch, out in the *Sirius*, and now I've got so I really like it. Here, we'd better keep in touch." He found her hand and tucked it under his arm. "Stabilize our positions more, besides keeping us from getting too lonesome, here in the dark," he concluded, in a matter-of-fact voice.

"Thanks for saying 'us'—but you would, wouldn't you?" and a wave of admiration went through her for the real and chivalrous manhood of the man with whom she had been forced by circumstances to cast her lot. "How long must we stay here?"

"As long as the air lasts, and I'd like to stay here longer than that. We don't want to move around any more than we absolutely have to until their rays are off of us, and we have no way of knowing how long that will be. Also, we'd better keep still. I don't know what kind of an audio system they've got, but there's no use taking unnecessary chances."

"All x—I'm an oyster's little sister," and for many minutes the two remained motionless and silent. Now and then Nadia twitched and started at some vague real or imaginary sound—now and then her fingers tightened upon his biceps—and he pressed her hand with his great arm in reassurance and understanding. Once a wall of their cell resounded under the impact of a fierce blow and Stevens instantly threw his arm around the girl, twisting himself between her and the threatened wall, ready for any emergency. But nothing more happened; the door remained closed, the cell stayed bottle-tight, and time wore slowly on. All too soon the unmistakable symptoms of breathing an unfit atmosphere made themselves apparent and Stevens, after testing each of the doors, drew the girl into a larger room, where they breathed deeply of the fresh, cool air.

"How did you know that this room was whole?" asked Nadia. "We might have stepped out into space, mightn't we?"

"No; if this room had lost its tightness, the door wouldn't have opened. They won't open if there's a difference of one kilogram pressure on the two sides. That's how I knew that the room we were in at first was cut in two—the door into that air-break wouldn't move."

"What comes next?"

"I don't know exactly what to do—we'd better hold a little council of war. They may have gone..." Stevens broke off as the structure began to move, and they settled down upon what had been one of the side-walls. Greater and greater became the acceleration, until their apparent weight was almost as much as it would have been upon the Earth, at which point it became constant. "... but they haven't," he continued the interrupted sentence. "This seems to be a capture and seizure, as well as an attack, so we'll have to take the risk of looking at them. Besides, it's getting cold in here. One or two of the adjoining cells have apparently been ruptured and we're radiating our heat out into space, so we'll have to get into a life-boat or freeze. I'll go pick out the best one. Wonder if I'd better take you with me, or hide you and come back after you?"

"Don't worry about that—I'm coming with you," Nadia declared, positively.

"Just as well, probably," he assented, and they set out. A thorough exploration of all the tight connecting cells revealed that not a lifeboat within their reach remained intact, but that habitable and navigable portions of three such craft were available. Selecting the most completely equipped of these, they took up their residence therein by entering it and closing the massive insulating door. Stevens disconnected all the lights save one, and so shielded that one before turning it on that it merely lightened the utter darkness into a semi-permeable gloom. He then stepped up to the lookout plate, and with his hand upon the control, pondered long the possible consequences of what he wished to do.

"What harm would it do to take just a little peek?"

"I don't know—that's the dickens of it. Maybe none, and then again, maybe a lot. You see, we don't know who or what we are up against. The only thing we know is that they've got us beat a hundred ways, and we've got to act accordingly. We've got to chance it sometime, though, if we can ever get away, so we might as well do it now. I'll put it on very short range first, and see what we can see. By the small number of cells we've got here I'm afraid they've split us up lengthwise, too—so that instead of having a whole slice of the old watermelon to live in, we've got only about a sixth of one—shaped about like a piece of restaurant pie. One thing I can do, though. I'll turn on the communicator receiver and put it on full coverage—maybe we can hear something useful."

Putting a little power upon the visiray plate, he moved the point of projection a short distance from their hiding-place, so that the plate showed a view of the wreckage. The upper half of the vessel was still intact, the lower half a jumble of sharply-cut fragments. From each of the larger pieces a brilliant ray of tangible force stretched outward. Suddenly their receiver sounded behind them, as the high-powered transmitter in the telegraph room tried to notify headquarters of their plight.

"*Arcturus* attacked and cut up being taken tow...."

Rapidly as the message was uttered the transmitter died with a rattle in the middle of a word, and Nadia looked at Stevens with foreboding in her eyes.

"They've got something, that's one thing sure, to be able to neutralize our communicator beams that way," he admitted. "Not so good—we'll have to play this close to our vests, girl!"

"Are you just trying to cheer me up, or do you really think we have a chance?" she demanded. "I want to know just where we stand."

"I'm coming clean with you, no kidding. If we can get away, we'll be all x, because I'll bet a farm that by this time Brandon's got everything those birds have, and maybe more. They beat us to it, that's all. I'm kind of afraid, though, that getting away isn't going to be quite as simple as shooting fish down a well."

FAR ahead of them a port opened, a lifeboat shot out at its full power, and again their receiver tried to burst into sound, but it was a vain attempt. The sound died before one complete word could be uttered, and the lifeboat, its power completely neutralized by the rays of the tiny craft of the enemy, floated gently back toward the mass of its parent and accompanied it in its headlong flight. Several more lifeboats made the attempt, as the courageous officers of the *Arcturus*, some of whom had apparently succeeded in eluding the vigilance of the captors, launched the little shells from various ports; but as each boat issued, its power was neutralized and it found itself dragged helplessly along in the grip of one of those mysterious, brilliant rays of force. At least one hidden officer must have been watching the fruitless efforts, for the next lifeboat to issue made no attempt, either to talk or to flee, but from it there flamed out into space a concentrated beam of destruction—the terrible ray of annihilation, against which no known substance could endure for a moment; the ray which had definitely outlawed war. But even that frightful weapon was useless—it spent its force harmlessly upon an impalpable, invisible barrier, a hundred yards from its source, and the bold lifeboat disappeared in one blinding explosion of incandescence as the captor showed its real power in retaliation. Stevens, jaw hard-set, leaped from the screen, then brought himself up so quickly that he skated across the smooth steel floor. Shutting off the lookout plate, he led the half-fainting girl across the room to a comfortable seat and sat down beside her—raging, but thoughtful. Nadia soon recovered.

"Why are you acting so contrary to your nature—is it because of *me*?" she demanded. "A dozen times I've seen you start to do something and then change your mind. I *will not* be a load on you nor hinder you in anything you want to do."

"I told your father I'd look after you, and I'm going to do it," he replied, indirectly. "I would do it anyway, of course—even if you are ten or twelve years older than I thought you were."

"Yes, Dad never has realized that I'm more than eight years old. I see—you were going out there and be slaughtered?" He flushed, but made no reply. "In that case I'm glad I'm here—that would have been silly. I think we'd better hold that council of war you mentioned a while ago, don't you?"

"I need a smoke—do you indulge?"

"No thanks. I tried it a few times at school, but never liked it."

He searched his pockets, bringing to light an unopened package and a tattered remnant which proved to contain one dilapidated cigarette. He studied it thoughtfully. "I'll smoke this wreck," he decided, "while it's still smokable. We'll save the rest of them—I'm afraid it'll be a long time between smokes. Well, let's confer!"

"This will have to be a one-sided conference. I don't imagine that any of my ideas will prove particularly helpful. You talk and I'll listen.

"You can't tell what ideas may be useful—chip in any time you feel the urge. Here's the dope, as I see it. They're highly intelligent creatures and are in all probability neither Martians nor Venerians. If any of them had any such stuff as that, some of us would have known about it and, besides, I don't believe they would have used it in just that way. Mercury is not habitable, at least for organic beings; and we have never seen any sign of any other kind of inhabitants who could work with metals and rays. They're probably from Jupiter, although possibly from further away. I say Jupiter, because I would think, judging from the small size of the ship, that it may still be in the experimental stage, so that they probably didn't come from any further away than Jupiter. Then, too, if they were very numerous, somebody would have sighted one before. I'd give my left leg and four fingers for one good look at the inside of that ship."

"Why didn't you take it, then? You never even looked toward it, after that one first glimpse."

"I'll say I didn't—the reason being that they may have automatic detectors, and as I have suggested before, our system of vision is so crude that its use could be detected with a clothesline or a basket full of scrap iron. But to resume: Their aim is to capture, not destroy, since they haven't killed anybody except the one crew that attacked them. Apparently they want to study us or something. However, they don't intend that any of us shall get away, nor even send out a word of what has happened to us. Therefore it looks as though our best bet is to hide now, and try to sneak away on them after a while—direct methods won't work. Right?"

"You sound lucid. Is there any possibility of getting back, though, if we got anywhere near Jupiter? It's so far away!"

"It's a long stretch from Jupiter to any of the planets where we have power-plants, all right—particularly now, when Mars and Tellus are subtending an angle of something more than ninety degrees at the sun, and Venus is between the two, while Jupiter is clear across the sun from all three of them. Even when Jupiter is in mean opposition to Mars, it is still some five hundred and fifty million kilometers away, so you can form some idea as to how far it is from our nearest planet now. No, if we expect to get back under our own power, we've got to break away pretty quick—these lifeboats have very little accumulator capacity, and the receptors are useless above about three hundred million kilometers...."

"But it'll take us a long time to go that far, won't it?"

"Not very. Our own ships, using only the acceleration of gravity, and both plus and minus at that, make the better than four hundred million kilometers of the long route to Mars in five days. These birds are using almost that much acceleration, and I don't see how they do it. They must have a tractor ray. Brandon claimed that such a thing was theoretically possible, but Westfall and I couldn't see it. We ragged him about it a lot—

and he was right. I thought, of course, they'd drift with us, but they are using power steadily. They've got *some* system!"

"Suppose they could be using intra-atomic energy? We were taught that it was impossible, but you've shattered a lot of my knowledge today."

"I wouldn't want to say definitely that it is absolutely impossible, but the deeper we go into that line, the more unlikely intra-atomic energy power-plants become. No, they've got a real power-transmission system—one that can hold a tight beam together a lot farther than anything we have been able to develop, that's all. Well, we've given them quite a lot of time to get over any suspicion of us, let's see if we can sneak away from them."

BY short and infrequent applications of power to the dirigible projectors of the life-boat, Stevens slowly shifted the position of the fragment which bore their craft until it was well clear of the other components of the mass of wreckage. He then exerted a very small retarding force, so that their bit would lag behind the procession, as though it had accidently been separated. But the crew of the captor was alert, and no sooner did a clear space show itself between them and the mass than a ray picked them up and herded them back into place. Stevens then nudged other pieces so that they fell out, only to see them also rounded up. Hour after hour he kept trying—doing nothing sufficiently energetic to create any suspicion, but attempting everything he could think of that offered any chance of escape from the clutches of their captors. Immovable at the plate, his hands upon the controls, he performed every insidious maneuver his agile brain could devise, but he could not succeed in separating their vehicle from its fellows. Finally, after a last attempt, which was foiled as easily as were its predecessors, he shut off his controls and turned to his companion with a grin.

"I didn't think I could get away with it—they're keen, that gang—but I had to keep at it as long as it would have done us any good."

"Wouldn't it do us any good now?"

"Not a bit—we're going so fast that we couldn't stop—we're out of even radio range of our closest power-plant. We'll have to put off any more attempts until they slow us down. They're fairly close to at least one of the moons of Jupiter, we'll have our best chance—so good, in fact, that I really think we can make it."

"But what good would that do us, if we couldn't get back?" Dire foreboding showed in her glorious eyes.

"Lots of things not tried yet, girl, and we'll try them all. First, we get away. Second, we try to get in touch with Norman Brandon...."

"How? No known radio will carry half that far."

"No, but I think that a radio as yet unknown may be able to—and there is a bare possibility that I'll be able to communicate."

"Oh wonderful—that lifts a frightful load off my mind," she breathed.

"But just a minute—I said I'd come clean with you, and I will. The odds are all against us, no matter what we do. If that unknown radio won't work—and it probably won't—there are several other things we can try, but they're all pretty slim chances. Even if we get away, it'll probably be about the same thing as though you were to be marooned on a desert island without any tools, and with your rescue depending upon your ability to build a high-powered radio station with which to call to a mainland for help. However, if we don't try to get away, our only alternative is letting them know we're here, and joining our friends in captivity."

"And then what?"

"You know as much as I do. Imprisonment and restraint, certain; death, possible; return to Earth, almost certainly impossible—life as guests, highly improbable."

"I'm with you, Steve, all the way."

"Well, it's time to spring off—we've both been awake better than fifty hours. Personally, I'm all in, and you're so near dead that you're a physical wreck. We'll get us a bite of supper and turn in."

An appetizing supper was prepared from the abundant stores and each ate a heartier meal than either would have believed possible. Stevens considered his unopened package of cigarettes, then regretfully put it back into his pocket still unopened and turned to Nadia.

"Well, little fellow, it's time to shove off, and then some. You might as well sleep here, and I'll go in there. If anything scares you, yell. Good-night, old trapper!"

"Wait a minute, Steve." Nadia flushed, and her brown eyes and black eyebrows, in comparison with her golden-blond hair, lent her face a quizzical, elfin expression that far belied her feelings as she stared straight into his eyes. "I've never even been away from the Earth before, and with all this happening I'm simply scared to death. I've been trying to hide it, but I couldn't stand it alone, and we're going to be together too long and too close for senseless conventions to affect us. There's two bunks over there—why don't you sleep in one of them?"

He returned her steadfast gaze for a moment in silence.

"All x with me, Nadia," he answered, keeping out of his voice all signs of the tenderness he felt for her, and of his very real admiration for her straightforward conduct in a terrifying situation. "You trust me, then?"

"*Trust* you! Don't be silly—I know you! I know you, and I know Brandon and Westfall—I know what you've done, and exactly the kind of men you are. *Trust* you!"

"Thanks, old golf-shootist," and promises were made and received in a clasp from which Nadia's right hand, strong as it was, emerged slightly damaged.

"By the way, what is your first name, fellow-traveller?" she asked in lighter vein. "Nobody, not even Dad or Breckie, ever seems to call you anything but 'Steve' when they talk about you." She was amazed at the effect of her innocent question, for Stevens flushed to his hair and spluttered.

"It's *Percy*!" He finally, snorted. "Percival Van Schravendyck Stevens. Wouldn't that tear it?"

"Why, I think Percival's a real nice name!"

"Silence!" he hissed in burlesque style. "Young woman, I have revealed to you a secret known to but few living creatures. On your life, keep it inviolate!"

"Oh, very well, if you insist. Good-night—Steve!" and she gave him a radiant and honest smile: the first smile he had seen since the moment of the attack.

CHAPTER III
Castaways Upon Ganymede

UPON awakening, the man's first care was to instruct the girl in the operation of the projectors, so that she could keep the heavily-armored edge of their small section, which she had promptly christened "The Forlorn Hope," between them and the grinding, clashing mass of wreckage, and thus, if it should become necessary, protect the relatively frail inner portions of their craft from damage.

"Keep an eye on things for a while, Nadia," he instructed, as soon as she could handle the controls, "and don't use any more power than is absolutely necessary. We'll need it all, and besides, they can probably detect anything we can use. There's probably enough leakage from the ruptured accumulator cells to mask quite a little emission, but don't use much. I'm going to see what I can do about making this whole wedge navigable."

"Why not just launch what's left of this lifeboat? It's space-worthy, isn't it?"

"Yes, but it's too small. Two or three of the big dirigible projectors of the lower band are on the rim of this piece-of-pie-shaped section we're riding, I think. If so, and if enough batteries of accumulators are left intact to give them anywhere nearly full power, we can get an acceleration that will make a lifeboat look sick. Those main dirigibles, you know, are able to swing the whole mass of the *Arcturus*, and what they'll do to this one chunk of it—we've got only a few thousand tons of mass in this piece—will be something pretty. Also, having the metal may save us months of time in mining it."

He found the projectors, repaired or cut out the damaged accumulator cells, and reconnected them through the controls of the lifeboat. He moved into the "engine-room" the airtanks, stores, and equipment from all the other fragments which, by means of a space-suit, he could reach without too much difficulty. From the battery rooms of those fragments—open shelves, after being sliced open by the shearing ray—he helped himself to banks of accumulator cells from the enormous driving batteries of the ill-fated *Arcturus,* bolting them down and connecting them solidly until almost every compartment of their craft was one mass of stored-up energy.

Days fled like hours, so furiously busy were they in preparing their peculiar vessel for a cruise of indefinite duration. Stevens cut himself short on sleep and snatched his meals in passing; and Nadia, when not busy at her own tasks of observing, housekeeping, and doing what little piloting was required, was rapidly learning to wield most effectively the spanner and pliers of the mechanic and electrician.

"I'm afraid our time is getting short, Steve," she announced, after making an observation. "It looks as though we're getting wherever it is we're going."

"Well, I've got only two more jobs to do, but they're the hardest of the lot. It is Jupiter, or can you tell yet?"

"Jupiter or one of its satellites, I think, from the point where they reversed their power. Here's the observation you told me to take."

"Looks like Jupiter," he agreed, after he had rapidly checked her figures. "We'll pass very close to one of those two satellites—probably Ganymede—which is fine for our scheme. All four of the major satellites have water and atmosphere, but Ganymede, being largest, is best for our purposes. We've got a couple of days yet—just about time to finish up. Let's get going—you know what to do."

"Steve, I'm afraid of it. It's too dangerous—isn't there some other way?"

"None that I can see. The close watch they're keeping on every bit of this junk makes it our only chance for a get-away. I'm pretty sure I can do it—but if I should happen to get nipped, just use enough power to let them know you're here, and you won't be any worse off than if I hadn't tried to pull off this stunt."

He donned a space-suit, filled a looped belt with tools, picked up a portable power-drill, and stepped into the tiny air-lock. Nadia deftly guided their segment against one of the larger fragments and held it there with a gentle, steady pressure, while Stevens, a light cable paying out behind him, clambered carefully over the wreckage, brought his drill into play, and disappeared inside the huge wedge. In less than an hour he returned without mishap and reported to the glowing girl.

"Just like shooting fish down a well! Most of the accumulator cells were tight, and installing the relays wasn't a bad job at all. Believe me, girl, there'll be junk filling all the space between here and Saturn when we touch them off!"

"Wonderful, Steve!" Nadia exclaimed. "It won't be so bad seeing you go into the others, now that you have this one all rigged up."

AROUND and around the mass of wreckage they crept, and in each of the larger sections Stevens connected up the enormous fixed or dirigible projectors to whatever accumulator cells were available through sensitive relays, all of which he could close by means of one radio impulse. The long and dangerous task done, he stood at the lookout plate, studying the huge disk which had been the upper portion of the lower half of the *Arcturus* and frowning in thought. Nadia reached over his shoulder and switched off the plate.

"Nix on that second job, big fellow!" she declared. "They aren't really necessary, and you're altogether too apt to be killed trying to get them. It's too ghastly—I won't stand for your trying it, so that ends it."

"We ought to have them, really," he protested. "With those special tools, cutting torches, and all the stuff, we'd be sitting pretty. We'll lose weeks of time by not having them."

"We'll just have to lose it, then. You can't get 'em, any more than a baby can get the moon, so stop crying about it," she went over the familiar argument for the twentieth time. "That stuff up there is all grinding together like cakes of ice in a floe; the particular section you want is in plain sight of whoever is on watch; and those tools and things are altogether too heavy to handle. You're a husky brute, I know, but even you couldn't begin to handle them, even if you had good going. I couldn't help you very much, even if you'd let me try; and the fact that you so positively refuse to let me come along shows how dangerous you know the attempt is bound to be. You'd probably never even get up there alive, to say nothing of getting back here. No, Steve, that's out like a light."

"I sure wish they'd left us weightless for a while, sometime, if only for an hour or two," he mourned.

"But they didn't!" she retorted, practically. "So we're just out of luck to that extent. Our time is about up, too. It's time you worked us back to the tail end of this procession—or rather, the head end, since we're traveling 'down' now."

Stevens took the controls and slowly worked along the outer edge of the mass, down toward its extremity. Nadia put one hand upon his shoulder and he glanced around.

"Thanks, Steve. We have a perfectly wonderful chance as it is, and we've gone so far with our scheme together that it would be a crying shame not to be able to go through with it. I'd hate like sin to have to surrender to them now, and that's all I could do if anything should become of you. Besides..." her voice died away into silence.

"Sure, you're right," he hastily replied, dodging the implication of that unfinished sentence. "I couldn't figure out anything that looked particularly feasible anyway—that's why I didn't try it. We'll pass it up."

Soon they arrived at their objective and maintained a position well in the van, but not sufficiently far ahead of the rest to call forth a restraining ray from their captors. Already strongly affected by the gravitational pull of the mass of the satellite, many of the smaller portions of the wreck, not directly held by the tractors, began to separate from the main mass. As each bit left its place another beam leaped out, until it became apparent that no more were available, and Stevens strapped the girl and himself down before two lookout plates.

"Now for it, Nadia!" he exclaimed, and simultaneously threw on the power of his own projectors and sent out the radio impulse which closed the relays he had so carefully set. They were thrown against the restraining straps savagely and held there by an enormous weight as the gigantic dirigible projectors shot their fragment of the wreck away from the comparatively slight force which had been acting upon it, but they braced themselves and strained their muscles in order to watch what was happening. As the relays in the various fragments closed, the massed power of the accumulators was shorted dead across the converters and projectors instead of being fed into them gradually through the controls of the pilot, with a result comparable to that of the explosion of an ammunition dump. Most of the masses, whose projectors were fed by comparatively few accumulator cells, darted away entirely with a stupendous acceleration. A few of them, however, received the unimpeded flow of complete batteries. Those projectors tore loose from even their massive supports and crashed through anything opposing them like a huge, armor-piercing projectile. It was a spectacle to stagger the imagination, and Stevens grinned as he turned to the girl, who was staring in wide-eyed amazement.

"Well, ace, I think they're busy enough now so that it'll be safe to take that long-wanted look at their controls," and he flashed the twin beams of his lookout light out beyond the upper half of the *Arcturus*—only to see them stop abruptly in mid-space. Even the extremely short carrier-wave of Roeser's Rays could not go through the invisible barrier thrown out by the tiny, but powerful globe of space.

"No penetration?" Nadia asked.

"Flattened them out cold. 'However,' as the fox once remarked about the grapes, 'I'll bet they're sour, anyway.' We'll have some stuff of our own, one of these days. I sure hope the fireworks we started back there keep those birds amused until we get out of sight, because if I use much more power on these projectors we may not have juice enough left to stop with."

"You're using enough now to suit me—I'm so heavy I can hardly lift a finger!"

"You'd better lift 'em! You must watch what's going on back there while I navigate around this moon."

"All x, chief.... They've got their hands full, apparently. Those rays are shooting around all over the sky. It looks as though they were trying to capture four or five things at once with each one."

"Good! Tell me when the moon cuts them off."

AT the awful acceleration they were using, which constantly increased the terrific velocity with which they had been traveling when they made good their escape, it was not long until they had placed the satellite between them and the enemy; then Stevens cut down and reversed his power. Such was their speed, however, that a long detour was necessary in order to reduce it to a safe landing rate. As soon as this could be done, Stevens headed for the morning zone and dropped the "Hope" rapidly toward the surface of that new, strange world. Details could not be distinguished at first because of an all-enshrouding layer of cloud, but the rising sun dispelled the mist, and when they had descended to within a few thousand feet of the surface, their vision was unobstructed. Immediately below them the terrain was mountainous and heavily wooded; while far to the east the rays of a small, pale sun glinted upon a vast body of water. No signs of habitation were visible as far as the eye could reach.

"Now to pick out a location for our power-plant. We must have a waterfall for power, a good place to hide our ship from observation, and I'd like to have a little seam of coal. We can use wood if we have to, but I think we can find some coal. This is all sedimentary rock—it looks a lot like the country along the North Fork of the Flathead, in Montana. There are a lot of coal outcrops, usually, in such topography as this is."

"We want to hide in a hurry, though, don't we?"

"Not particularly, I think. If they had missed us at all, they would have had us long ago, and with all the damage we did with those projectors they won't be surprised at one piece being missing—I imagine they lost a good many."

"But they'll know that somebody caused all that disturbance. Won't they hunt for us?"

"Maybe, and maybe not—no telling what they'll do. However, by the time they can land and get checked up and ready to hunt for us, we'll be a mighty small needle, well hidden in a good big haystack."

For several hours they roamed over the mountainous region at high velocity, seeking the best possible location, and finally they found one that was almost ideal—a narrow canyon overhung with heavy trees, opening into a wide, deep gorge upon a level with its floor. A mighty waterfall cascaded into the gorge just above the canyon, and here and there could be seen black outcrops which Stevens, after a close scrutiny, declared to be coal. He deftly guided their cumbersome wedge of steel into the retreat, allowed it to settle gently to the ground, and shut off the power.

"Well, little fellow-conspirator against the peace and dignity of the Jovians, I don't know just where we are, but wherever it is, we're here. We got away clean, and as long as we don't use any high-tension stuff or anything else that they can trace, I think we're as safe as money in a bank."

"I suppose that I ought to be scared to death, Steve, but I'm not—I'm just too thrilled for words," Nadia answered, and the eager sparkle in her eyes bore out her words. "Can we go out now? How about air? Shall we wear suits or go out as we are? Have you got a weapon of any kind? Hurry up—let's do something!"

"Pipe down, ace! Remember that we don't know any more about anything around here than a pig does about Sunday, and conduct yourself accordingly. Take it easy. I'm surprised at the gravity here. This is certainly Ganymede, and it has a diameter of only about fifty seven hundred kilometers. If I remember correctly, Damoiseau estimated its mass at about three one-hundredths that of the Earth, which would make its surface gravity about one-sixth. However, it is actually almost a half, as you see by this spring-balance here. Therefore it is quite a little more massive than has been...."

"What of it? Let's go places and do things!"

"Calm yourself, Ginger, you've got lots of time—we'll be here for quite a while, I'm afraid. We can't go out until we analyze the air—we're sure lucky there's as much as there is. I'm not exactly the world's foremost chemist, but fortunately an air-analysis isn't much of a job with the apparatus we carry."

While Nadia controlled her impatience as best she could, Stevens manipulated the bulbs and pipettes of the gas apparatus.

"Pressure, fifty-two centimeters—more than I dared hope for—and analysis all x, I believe. Oxygen concentration a little high, but not much."

"We won't have to wear the space-suits, then?"

"Not unless I missed something in the analysis. The pressure corresponds to our own at a height of about three thousand meters, which we can get used to without too much trouble. Good thing, too. I brought along all the air I could get hold of, but as I told you back there, if we had to depend on it altogether, we might be out of luck. I'm going to pump some of our air back into a cylinder to equalize our pressure—don't want to waste any of it until we're sure the outside air suits us without treatment."

WHEN the pressure inside had been gradually reduced to that outside and they had become accustomed to breathing the rarefied medium, Stevens opened the airlock and the outside doors, and for some time cautiously sniffed the atmosphere of the satellite. He could detect nothing harmful or unusual in it—it was apparently the same as earthly air—and he became jubilant.

"All x, Nadia—luck is perched right on our banner. Freedom, air, water, power, and coal! Now as you suggested, we'll go places and do things!"

"Suppose it's safe?" Her first eagerness to explore their surroundings had abated noticeably. "You aren't armed, are you?"

"No, and I don't believe that there was a gun of any kind aboard the *Arcturus*. That kind of thing went out quite a while ago, you know. We'll take a look, anyway—we've got to find out about that coal before we decide to settle down here. Remember this half-gravity stuff, and control your leg-muscles accordingly."

Leaping lightly to the ground, they saw that the severed section of fifty-inch armor, which was the rim of their conveyance, almost blocked the entrance to the narrow canyon which they had selected for their retreat. Upon one side that wall of steel actually touched the almost perpendicular wall or rock; upon the other side there was left only a narrow passage. They stepped through it, so that they could see the waterfall and the gorge, and stopped silent. The sun, now fairly high, was in no sense the familiar orb of day, but was a pale, insipid thing, only one-fifth the diameter of the sun to which they were accustomed, and which could almost be studied with the unshielded eye. From their feet a grassy meadow a few hundred feet wide sloped gently down to the river, from whose farther bank a precipice sprang upward for perhaps a thousand feet—merging into towering hills whose rugged grandeur was reminiscent of the topography of the moon. At their backs the wall of the gorge was steep, but not precipitous, and was covered with shrubs and trees—some of which leaned out over the little canyon, completely screening it, and among whose branches birds could now and then be seen flitting about. In that direction no mountains were visible, indicating that upon their side of the river there was an upland plateau or bench. To their right the river, the gorge, and the strip of meadow extended for a mile or more, then curved away and were lost to sight. To their left, almost too close for comfort, was the stupendous cataract, towering above them to a terror-inspiring height. Nadia studied it with awe, which changed to puzzled wonder.

"What's the matter with it, Steve? It looks like a picture in slow motion, like the kind they take of your dives—or am I seeing things?"

"No, it's really slow, compared to what we're used to. Remember that one-half gravity stuff!"

"Oh, that's right, but it certainly does look funny. It gives me the creeps."

"You'll get used to it pretty quick—just as you'll get used to all the rest of the things having only half their earthly weight and falling only half as fast as they ought to when you drop them. Well, I don't see anything that looks dangerous yet—let's go up toward the falls a few meters and prospect that outcrop."

With a few brisk strokes of an improvised shovel he cleared the outcrop of detritus and broke off several samples of the black substance, with which they went back to the "Forlorn Hope."

"It's real coal," Stevens announced after a series of tests. "I've seen better, but on the other hand, there's lots worse. It'll make good gas, and a kind of a coke. Not so hot, but it'll do. Now we'd better get organized old partner, for a long campaign."

"Go ahead and organize—I'm only the cheap help in this enterprise."

"Cheap help! You're apt to be the life of the party. Can you make and shoot a bow and arrow?"

"I'll say I can—I've belonged to an archery club for five years."

"What did I tell you? You're a life saver! Here's the dope—we've got to save our own supplies as much as possible until we know exactly what we're up against, and to do that, we've got to live off the country. I'll fake up something to knock over some of those birds and small game, then we can make real bow-strings and feathered arrows and I'll forge some steel arrow-heads while you're making yourself a real bow. We'd better make me about a hundred-pound war bow, too...."

"A *hundred*!" interrupted Nadia. "That's a lot of bow, big boy—think you can bend it?"

"You'd be surprised," he grinned. "I'm not quite like Robin Hood—I've been known to miss a finger-thick wand at a hundred paces—but I'm not exactly a beginner."

"Oh, of course—I should have known by your language that you're an archer, otherwise you'd never have used such an old-fashioned word as 'pounds.' I shoot a thirty-five-pound bow ordinarily, but for game I should have the heaviest one I can hold accurately—about a forty-five, probably."

"All x. And as soon as I can I'll make us a couple of suits of fairly heavy steel armor, so that we'll have real protection if we should need it. You see, we don't know what we are apt to run up against out here. Then, with that much done, it'll be up to you to provide, since I'll have to work tooth and nail at the forges. You'll have to bring home the bacon, do the cooking and so on, and see what you can find along the line of edible roots, grains, fruits, and what-not. Sort of reverse the Indian idea—you be the hunter and I'll keep the home fires burning. Can do?"

"What it takes to do that, I've got," Nadia assured him, her eyes sparkling. "Have you your job planned out as well and as fittingly as you have mine?"

"And then some. We've got just two methods of getting away from here—one is to get in touch with Brandon, so that he'll come after us; the other is to recharge our accumulators and try to make it under our own power. Either course will need power and lots of it...."

"I never thought of going back in the 'Hope.' Suppose we could?"

"About as doubtful as the radio—I think that I could build a pair of matched-frequency auto-dirigible transmitter and receptor units, such as are necessary for space-ships fed by stationary power-plants, but after I got them built, they'd take us less than half way there. Then we'd have only what power we can carry, and I hate even to think of what probably would happen to us. We'd certainly have to drift for months before we could get close enough to any of our plants to radio for help, and we'd be taking awful chances. You see, we'd have to take a very peculiar orbit, and if we should miss connections passing the inner planets, what the sun would do to us at the closest point and where what's left of us would go on the back-swing, would be just too bad! Besides, if we can get hold of the *Sirius*, they'll come loaded for bear, and we may be able to do something about the rest of the folks out here."

"OH!" breathed the girl. "Wouldn't it be wonderful if we could! I thought, of course, they'd all be...." her voice died away.

"Not necessarily—there's always a chance. That's why I'm trying the ultra-radio first. However, either course will take lots of power, so the first thing I've got to do is to build a power plant. I'm going to run a penstock up those falls, and put in a turbine, driving a high-tension alternator. Then, while I'm trying to build the ultra-radio, I'll be charging our accumulators, so that no time will be lost in case the radio fails. If it does fail—and remember I'm not counting on its working—of course I'll tackle the transmission and receptor units before we start out to drift it."

"You say it easy, Steve, but how can you build all those things, with nothing to work with?"

"It's going to be a real job—I'm not try to kid you into thinking it'll be either easy or quick. Here's the way everything will go. Before I can even lay the first length of the penstock, I've got to have the pipe—to make which I've got to have flat steel—to get which I'll have to cut some of the partitions out of this ship of ours—to do which I'll have to have a cutting torch—to make which I'll have to forge nozzles out of block metal and to run which I'll have to have gas—to get which I'll have to mine coal and build a gas-plant—to do which...."

"Good heavens, Steve, are you going back to the Stone Age? I never thought of half those things. Why, it's impossible!"

"Not quite, guy. Things could be a lot worse—that's why I brought along the whole 'Forlorn Hope,' instead of just the lifeboat. As it is, we've got several thousand tons of spare steel and lots of copper. We've got ordinary tools and a few light motors, blowers, and such stuff. That gives me a great big start—I won't have to mine the ores and smelt the metals, as would have been necessary otherwise. However, it'll be plenty bad. I'll have to start out in a pretty crude fashion, and for some of the stuff I'll need I'll have to make, not only the machine that makes the part I want, but also the machine that makes the machine that makes the machine that makes it—and so on, just how far down the line, I haven't dared to think."

"You must be a regular jack-of-all-trades, to think you can get away with such a program as that?"

"I am—nothing else but. You see, while most of my school training was in advanced physics and mathematics, I worked my way through by computing and designing, and I've done a lot of truck-horse labor of various kinds besides. I can calculate and design almost anything, and I can make a pretty good stab at translating a design into fabricated material. I wouldn't wonder if Brandon's ultra-radio would stop me, since nobody had even started to build one when I saw him last—but I helped compute it, know the forces involved as well as he did at that time, and it so happens that I know more about the design of coils and fields of force than I do about anything else. So I may be able to work it out eventually. It isn't going to be not knowing how that will hold me up—it'll be the lack of something that I can't build."

"And that's where you will go back and back and back, as you said about building the penstock?"

"Back and back is right, if I can find all the necessary raw materials—that's what's probably going to put a lot of monkey-wrenches into the machinery." And Stevens went to work upon a weapon of offense, fashioning a crude, but powerful bow from a strip of spring steel strung with heavy wire.

"How about arrows? Shall I go see if I can hit a bird with a rock, for feathers, and see if I can find something to make arrows out of?"

"Not yet—anyway, I'd bet on the birds! I'm going to use pieces of this light brace-rod off the accumulator cells for arrows. They won't fly true, of course, but with their mass I can give them enough projectile force to kill any small animal they hit, no matter how they hit it."

After many misses, he finally bagged a small animal, something like a rabbit and something like a kangaroo, and a couple of round-bodied, plump birds, almost as large as domestic hens. These they dressed, with considerable distaste and a noticeable lack of skill.

"We'll get used to it pretty quick, Diana—also more expert," he said when the task was done. "We now have raw material for bow-strings and clothes, as well as food."

"The word 'raw' being heavily accented," Nadia declared, with a grimace. "But how do we know that they're good to eat?"

"We'll have to eat 'em and see," he grinned. "I don't imagine that any flesh is really poisonous, and we'll have to arrive at the ones we like best by a process of trial and error. Well, here's your job—I'll get busy on mine. Don't go more than a few hundred meters away and yell if you get into a jam."

"There's a couple of questions I want to ask you. What makes it so warm here, when the sun's so far away and Jupiter isn't supposed to be radiating any heat? And how about time? It's twelve hours by my watch since sunrise this morning, and it's still shining."

"As for heat, I've been wondering about that. It must be due to internal heat, because even though Jupiter may be warm, or even hot, it certainly isn't radiating much, since it has a temperature of minus two hundred at the visible surface, which, of course, is the top of the atmosphere. Our heat here is probably caused by radioactivity—that's the most modern dope, I believe. As for time, it looks as though our days were something better than thirty hours long, instead of twenty-four. Of course I'll keep the chronometer going on I-P time, since we'll probably need it in working out observations; but we might as well let our watches run down and work, eat, and sleep by the sun—not much sense in trying to keep Tellurian time here, as I see it. Check?"

"All x. I'll have supper ready for you at sunset. 'Bye!"

A few evenings later, when Stevens came in after his long day's work, he was surprised to see Nadia dressed in a suit of brown coveralls and high-laced moccasins.

"How do I look?" she asked, pirouetting gayly.

"Neat, but not gaudy," he approved. "That's good mole-skin—smooth, soft, and tough. Where'd you make the raise? I didn't know we had anything like that on board. What did you do for thread? You look like a million dollars—you sure did a good job of fitting."

"I had to have something—what with all the thorns and brush, there was almost more of me exposed than covered, and I was getting scratched up something fierce. So I ripped up one of the space-suits, and found out that there's enough cloth, fur, and leather in one of them to make six ordinary suits, and thread by the kilometer. I was awfully glad to see all that thread—I had an idea that I'd have to unravel my stockings or something, but I didn't. Your clothes are getting pretty tacky, too, and you're getting all burned with those hot coals and things. I'm going to build you a suit out of leather for your blacksmithing activities."

"Fine business, ace! Then we can save what's left of our civilized clothes for the return trip. What do we eat?"

"The eternal question of the hungry laboring man! I've got a roasted bongo, a fried filamaloo bird, and a boiled warple for the meat dishes. For vegetables, mashed hikoderms and pimola greens. Neocorn bread."

"Translate that, please, into terms of food."

"Translate it yourself, after you eat it. I changed the system on you today. I've named all the things, so it'll be easier to keep track of those we like and the ones we don't."

With appetites sharp-set by long hours of hard labor they ate heartily; then, in the deepening twilight, they sat and talked in comradely fashion while Stevens smoked one precious cigarette.

IT was not long until Nadia had her work well in hand. Game was plentiful, and the fertile valley and the neighboring upland yielded peculiar, but savory vegetable foods in variety and abundance; so that soon she was able to spend some time with Stevens, helping him as much as she could.

Thus she came to realize the true magnitude of the task he faced and the real seriousness of their position.

As Stevens had admitted before the work was started, he had known that he had set himself a gigantic task, but he had not permitted himself to follow, step by step, the difficulties that he knew awaited him. Now, as the days stretched into weeks and on into months, he was forced to take every laborious step, and it was borne in upon him just how nearly impossible that Herculean labor was to prove—just how dependent any given earthly activity is upon a vast number of others. Here he was alone—everything he needed must be manufactured by his own hands, from its original sources. He had known that progress would be slow and he had been prepared for that; but he had not pictured, even to himself, half of the maddening setbacks which occurred time after time because of the crudity of the tools and equipment he was forced to use. All too often a machine or part, the product of many hours of grueling labor, would fail because of the lack of some insignificant thing—some item so common as to be taken for granted in all terrestrial shops, but impossible of fabrication with the means at his disposal. At such times he would set his grim jaw a trifle harder, go back one step farther toward the Stone Age, and begin all over again—to find the necessary raw material or a possible substitute, and then to build the apparatus and machinery necessary to produce the part he required. Thus the heart-breaking task progressed, and Nadia watched her co-laborer become leaner and harder and more desperate day by day, unable in any way to lighten his fearful load.

In the brief period of rest following a noonday meal, Stevens lay prone upon the warm, fragrant grass beside the "Forlorn Hope," but it was evident to Nadia that he was not resting. His burned and blistered hands were locked savagely behind his head, his eyes were closed too tightly, and every tense line of his body was eloquent of a strain even more mental than physical. She studied him for minutes, her fine eyes clouded, then sat down beside him and put her hand upon his shoulder.

"I want to talk to you a minute, Steve," she said gently.

"All x, little fellow—but it might be just as well if you didn't touch me. You see, I'm getting so rabid that I can't trust myself."

"That's exactly what I want to talk to you about." A fiery blush burned through her deep tan, but her low, clear voice did not falter and her eyes held his unflinchingly. "I know you better than you know yourself, as I've said before. You are killing yourself, but it isn't the work, frightfully hard and disheartening as it is, that is doing it—it's your anxiety for me and the uncertainty of everything. You haven't been able to rest because you have been raging and fuming so at unavoidable conditions—you have been fighting *facts*. And it's all *so* useless, Steve, between you and me—everything would check out on zero if we'd just come out into the open."

The man's gaunt frame seemed to stiffen even more rigidly.

"You've said altogether too much or else only half enough, Nadia. You know, of course, that I've loved you ever since I got really to know you—and that didn't take long. You know that I love you and you know how I love you—with the real love that a man can feel for only one woman and only once in his life; and you know exactly what we're up against. Now that *does* tear it—wide open!" he finished bitterly.

"No, it doesn't, at all," she replied, steadily. "Of course I know that you love me, and I glory in it; and since you don't seem to realize that I love you in exactly the same way, I'll tell you so. Love you! Good heavens, Steve, I never dreamed that such a man as you are really existed! But you're fighting too many things at once, and they're killing you. And they're mostly imaginary, at that. Can't you see that there's no need of uncertainty between you and me? That there is no need of you driving yourself to desperation on my account? Whatever must be is all x with me, Steve. If you can build everything you need, all well and good. We'll be engaged until then, and our love will be open and sweet. If worst comes to worst, so that we can neither communicate with Brandon and Westfall nor leave here under our own power—even that is nothing to kill ourselves about. And yes, I do know exactly what we are facing. I have been prepared for it ever since I first saw what a perfectly impossible thing you are attempting. You are trying to go from almost the Age of Bronze clear up to year-after-next in a month or two. Not one man in a million could have done as much in his lifetime as you have done in the last few weeks, and I do not see how even you, with what little you have to work with, can possibly build such things as power-plants, transmitters, and ultra-radio stations. But what of it? For the day that it becomes clear that we are to remain here indefinitely; that day we will marry each other here, before God. Look around at this beautiful country. Could there be a finer world upon which to found a new race? When we decided to cut loose from the *Arcturus* I told you that I was with you all the way, and now I'll repeat it, with a lot more meaning. No matter what it's like, Steve, no matter where it leads to, I'm with you—*to—the—end—of—the—road*. Here or upon Earth or anywhere in the Universe. I am yours for life and for eternity."

WHILE she was speaking, the grim, strained lines upon Stevens' face had disappeared, and as she fell silent he straightened up and gently, tenderly, reverently he took her lithe body into his arms.

"You're right, sweetheart—everything *will* check out on zero, to nineteen decimals." He was a man transfigured. "I've been fighting windmills and I've been scared sick—but how was I to think that a wonder-girl like you could ever love a mutt like me? You certainly are the gamest little partner a man ever had. You're the world's straightest shooter, ace—you're a square brick if there ever was one. Your sheer nerve in being willing to go the whole route makes me love you more than ever, if such a thing can be possible, and it certainly puts a new face on the whole cock-eyed Universe for me. However, I don't believe it will come to that. After what you've just said, I sure will lick that job, regardless of how many different factories it takes to make one armature—I'll show that mess of scrap-iron what kind of trees make shingles!"

The girl still in his arms, he rose to his feet and released her slowly, reluctantly, unwilling ever to let her go. Then he shook himself, as though an overwhelming burden had been lifted from his shoulders, and laughed happily.

"See this cigarette?" he went on lightly. "The Last of the Mohicans. I'm going to smoke it in honor of our engagement." He drew the fragrant smoke deep into his lungs and frowned at her in mock seriousness.

"This would be a nice world to live on, of course, but the jobs here are too darn steady. It also seems to be somewhat lacking in modern conveniences, such as steel-mills and machine tools. Then, too, it is just a trifle too far from the Royal and Ancient for you really to enjoy living here permanently, and besides, I can't get my favorite brand of cigarettes around here. Therefore, after due deliberation, I don't believe we'll take the place—we'll go back to Tellus. Kiss me just once more ace, and I'll make that job think a cyclone has struck it right on the center of impact. Like Samuel Weller, or whoever it was, I'm clear full of 'wigor, wim, and witality'!"

The specified kiss and several others duly delivered he strode blithely away, and the little canyon resounded with the blows of his heavy sledge as he attacked with renewed spirit the great forging, white-hot from his soak-pit, which was to become the shaft of his turbo-alternator. Nadia watched him for a moment, her very heart in her eyes, then picked up her spanner and went after more steel, breathing a long and tremulous, but supremely happy sigh.

CHAPTER IV
Ganymedean Life

SLOW, hard, and disheartening as the work had been at first, Stevens had never slackened his pace, and after a time, as his facilities increased, the exasperating setbacks decreased in number and severity and his progress became faster and faster. Large as the "Forlorn Hope" was, space was soon at a premium, for their peculiarly-shaped craft became a veritable factory, housing a variety of machinery and equipment unknown in any single earthly industrial plant. Nothing was ornamental—everything was stripped to its barest fundamental necessities—but every working part functioned with a smooth precision to delight the senses of any good mechanic.

In a cavern under the falls was the great turbine, to be full-fed by the crude but tight penstock which clung to the wall of the gorge, angling up to the brink of that stupendous cataract. Bedded down upon solid rock there was a high-tension alternator capable of absorbing the entire output of the mighty turbine. This turbo-alternator was connected to a set of converters from which the energy would flow along three great copper cables—the receptors of the lifeboats being altogether too small to carry the load—to the now completely exhausted accumulators of the "Forlorn Hope." All high-tension apparatus was shielded and grounded, so that no stray impulses could reveal to the possible detectors of the Jovians the presence of this foreign power plant. Housings, frames, spiders, all stationary parts were rough, crude and massive; but bearings, shafts, armatures, all moving parts, were of a polished and finished accuracy and balance that promised months and years of trouble-free operation. Everything ready for the test, Stevens took off his frayed and torn leather coveralls and moccasins and climbed nimbly up the penstock. He never walked down. Opening the head-gate, he poised sharply upon its extremity and took off in a perfect swan-dive; floating unconcernedly down toward that boiling maelstrom two hundred feel below. He struck the water with a sharp, smooth "slup!" and raced ashore, seizing his suit as he ran toward the turbo-alternator. It was running smoothly, and, knowing that everything was tight at the receiving end, he lingered about the power plant until he was assured that nothing would go wrong and that his home manufactured lubricating oil and grease would keep those massive bearings cool.

Hunger assailed him, and glancing at the sun, he noted that it was well past dinner-time.

"Wow!" he exclaimed aloud. "The boss just loves to wait meals—she'll burn me up for this!"

He ran lightly toward "home," eager to tell his sweetheart that the long awaited moment had arrived—that power was now flowing into their accumulators.

"Hi, Diana of the silver bow!" he called. "How come you no blow the dinner bell? Power's on—come give it a look!"

There was no answer to his hail, and Stevens paused in shocked amazement. He knew that never of her own volition would she be out so late—Nadia was gone! A rapid tour of inspection quickly confirmed that which he already knew only too well. Forgotten was his hunger, forgotten the power plant, forgotten everything except the fact that his Nadia, the buoyant spirit in whom centered his Universe, was lost or ... he could not complete the thought, even to himself.

Swiftly he came to a decision and threw off his suit, revealing the body of a Hercules—a body ready for any demand he could put upon it. Always in hard training, months of grinding physical labor and of heavy eating had built him up to a point at which he would scarcely have recognized himself, could he have glanced into a mirror. Mighty but pliable muscles writhed and swelled under his clear skin as he darted here and there, selecting equipment for what lay ahead of him. He donned the heavily armored space-suit which they had prepared months before, while they were still suspicious of possible attack. It was covered with heavy steel at every point, and the lenses of the helmet, already of unbreakable glass, had been re-enforced with thick steel bars. Tank and valves supplied air at normal pressure, so that his powerful body could function at full efficiency, not handicapped by the lighter atmosphere of Ganymede. The sleeves terminated in steel-protected rubber wristlets which left his hands free, yet sheltered from attack—wristlets tight enough to maintain the difference in pressure, yet not tight enough to cut off the circulation. He took up his mighty war-bow and the full quiver of heavy arrows—full-feathered and pointed with savagely barbed, tearing heads of forged steel—and slipped into their sheaths the long and heavy razor-sharp sword and the double-edged dirk, which he had made and ground long since for he knew not what emergency, and whose bell-shaped hilts of steel further protected his hands and wrists. Thus equipped, he had approximately his normal earthly weight; a fact which would operate to his advantage, rather than otherwise, in case of possible combat. With one last look around the "Forlorn Hope," whose every fitting spoke to him of the beloved mistress who was gone, he filled a container with water and cooked food and opened the door.

"IT won't be long now; now it won't be long." Nadia caroled happily, buckling on her pack straps and taking up bow and arrows for her daily hunt. "I never thought that he could do it, but what it takes to do things, he's got lots of," she continued to improvise the song as she left the "Hope" with its multitudinous devices whose very variety was a never-failing delight to her; showing as it did the sheer ability of the man, whose brain and hands had almost finished a next-to-impossible task.

Through the canyon and up a well-worn trail she climbed, and soon came out upon the sparsely timbered bench that was her hunting grounds. Upon this day, however, she was full of happy anticipation and her mind was everywhere except upon her work. She was thinking of Stevens, of their love, of the power which he might turn on that very day, and of the possible rescue for which she had hitherto scarcely dared to hope. Thus it was that she walked miles beyond her usual limits without having loosed an arrow, and she was surprised when she glanced up at the sun to see that half the morning was gone and that she was almost to the foothills, beyond which rose a towering range of mountains.

"Snap out of it, girl!" she reprimanded herself. "Go on wool-gathering like this and your man will go hungry—and he'll break you right off at the ankles!" She became again the huntress, and soon saw an animal browsing steadily along the base of a hill. It was a six-legged, deer-like creature, much larger than anything she had as yet seen. But it was meat and her time was short, therefore she crept within range and loosed an arrow with the full power of her hunting bow. Unfamiliar as she was with the anatomy of the peculiar creature, the arrow did not kill. The "hexaped," as she instantly named it, sped away and she leaped after it. She, like her companion, had developed amazingly in musculature, and few indeed were the denizens of Ganymede, who could equal her speed upon that small globe, with its feeble gravitational force.

Up the foothills it darted. Beyond the hills and deep into a valley between two towering peaks the chase continued before Nadia's third arrow brought the animal down. Bending over the game, she became conscious of a strange but wonderful sweet perfume and glanced up, to see something which she certainly had not noticed when the hexaped had fallen. It was an enormous flower, at least a foot in diameter and indescribably beautiful in its crimson and golden splendor. Almost level with her head the gorgeous blossom waved upon its heavy stem; based by a massive cluster of enormous, smooth, dark green leaves. Entranced by this unexpected and marvelous floral display, Nadia breathed deeply of the inviting fragrance—and collapsed senseless upon the ground. Thereupon the weird plant moved over toward her, and the thick leaves began to enfold her knees. This carnivorous thing, however, did not like the heavy cloth of her suit and turned to the hexaped. It thrust several of its leaves into the wounds upon the carcass and fed, while two other leaves rasped together, sending out a piercing call.

In answer to the sound the underbrush crackled, and through it and upon the scene there crashed a vegetable-animal nightmare—the parent of the relatively tiny thing whose perfume had disabled the girl.

Its huge and gorgeous blossom was supported by a long, flexible, writhing stem, and its base was composed of many and highly specialized leaves. There were saws and spears and mighty, but sinuous tendrils; there were slender shoots which seemed to possess some sense of perception; there was the massive tractor base composed of extensible leaves which by their contraction and expansion propelled the mass along the ground. Parent and child fell upon the hexaped and soon bones and hair were all that remained The slender shoots then wandered about the unconscious girl in her strange covering, and as a couple of powerful tendrils coiled about her and raised her into the air over the monstrous base of the thing, its rudimentary brain could almost be perceived working as it sluggishly realized that, now full fed, it should carry this other victim along, to feed its other offspring when they should return to its side.

BARELY outside the door of the "Forlorn Hope" Stevens whirled about with a bitter imprecation. He had already lost time needlessly—with a lookout plate he could cover more ground in ten minutes than he could cover afoot in a week. He flipped on the power and shot the violet beam out over the plateau to the district where he knew Nadia was wont to hunt. Not finding her there, he swung the beam in an ever widening circle around that district. Finally he saw a few freshly broken twigs, and scanned the scene with care. He soon found the trail of fresh blood which marked the path of the flight of the hexaped, and with the peculiar maneuverability of the device he was using, it was not long until he was studying the scene where the encounter had taken place. He gasped when he saw the bones and perceived three of Nadia's arrows, but soon saw that the skeleton was not human and was reassured. Casting about in every direction, he found Nadia's bow, and saw a peculiar, freshly trampled path leading from the kill, past the bow, down the valley. He could not understand the spoor, but it was easily followed, and he shot the beam along it at headlong speed until he came up with the monstrous creature that was making it—until he saw what burden that organism was carrying.

He leaped to the controls of the lifeboat, then dropped his hand. While the stream of power now flowing was ample to operate the lookout plates, yet it would be many hours before the accumulator cells would be in condition to drive the craft even that short distance.

"It'll take over an hour to get there—here's hoping I can check in all x," he muttered savagely, as he took careful note of the location and direction of the creature's trail and set off at a fast jog-trot.

The carnivorous flower's first warning that all was not well was received when Stevens' steel-shod feet landed squarely upon its base and one sweeping cut of his sword lopped off the malignant blossom and severed the two tendrils that still held the unconscious Nadia. With a quick heave of his shoulder, he tossed her lightly backward into the smooth-beaten track the creature had made and tried to leap away—but the instant he had consumed in rescuing the girl had been enough for the thing to seize him, and he found himself battling for his very life. No soft-leaved infant this, but a full-grown monster, well equipped with mighty weapons of offense and defense. Well it was for the struggling man that he was encased in armor steel as those saw-edged, hard-spiked leaves drove against him with crushing force; well it was for him that he had his own independent air supply, so that that deadly perfume eddied ineffective about his helmeted head! Hard and fiercely driven as those terrible thorns were, they could do no more than dent his heavy armor. His powerful left arm, driving the double-razor-edged dirk in short, resistless arcs, managed to keep the snaky tendrils from coiling about his right arm, which was wielding the heavy, trenchant sword. Every time that mighty blade descended it cleaved its length through snapping spikes and impotently grinding leaves; but more than once a flailing tendril coiled about his neck armor and held his helmet immovable as though in a vise, while those frightful, grinding saws sought to rip their way through the glass to the living creature inside the peculiar metal housing. Dirk and saber and magnificent physique finally triumphed, but it was not until each leaf was literally severed from every other leaf that the outlandish organism gave up the ghost.

NADIA had been tossed out into pure air, beyond the zone of the stupefying perfume, and she recovered her senses in time to see the finish of the battle. Stevens, assured that his foe was *hors du combat*, turned toward the spot where he had thrown Nadia's body. He saw that she was unharmed, and sprang toward her in relief. He was surprised beyond measure, however, to see her run away at a pace he could not hope to equal, encumbered as he was; motioning frantically at him the while to keep away from her. He stopped, astounded, and started to unscrew his helmet, whereupon she dashed back toward him, signaling him emphatically to leave his armor exactly as it was. He stood still and stared at her, an exasperated question large upon his face, until she made clear to him that he was to follow her at a safe distance, then she set off at a rapid walk. She led him back to where the hexaped had fallen, where she retrieved her bow and arrows; then, keeping a sharp lookout upon all sides, she went on to a small stream of water. She made the dumbfounded man go out into the middle of the creek and lie down and roll over in the water, approaching him sniffing cautiously between immersions. She made him continue the bathing until she could detect not even the slightest trace of the sweet, but noxious fragrance of that peculiarly terrible form of Ganymedean life. Only then did she allow him to remove his helmet, so that she could give him the greeting for which they both had longed and tell him what it was all about.

"So that's it, ace!" he exclaimed, still holding her tightly in his iron embrace. "Great balls of fire! I thought maybe you were still a little cuckoo. Anaesthetic perfume, huh? Hot stuff, I'd say—no wonder you bit—I would, too. It's lucky for us I was air-tight—we'd both be fee...."

"Stop it!" she interrupted him sharply, "Forget it—don't ever even think of it!"

"All x, ace. It's out like the well known light. What to do? It's getting darker than a hat, and we're a long way from home. Don't know whether I could find my way back in the dark or not; and just between you and me, I'm not particularly keen on night travel in these parts after what's just happened. Are you?"

"Anything else but," she assured him, fervently. "I'd lots rather stay hungry until tomorrow."

"No need of that—I've brought along enough supper for both of us. I'm hungry as a wolf, too, now that I have time to think of it. We'll eat and den up somewhere—or climb a tree. Those wampuses probably can't climb trees!"

"There's a nice little cave back there about a hundred meters. We'll pretend it's the Ritz," and they soon had a merry fire blazing in front of the retreat. There they ate of the provisions Stevens had brought. Then, while the man rolled up boulders before the narrow entrance of the cave, Nadia gathered leaves and made a soft bed upon its warm, dry floor.

"Good night, lover," and the girl, untroubled and secure now that Stevens was at her side, was almost instantly asleep; but the man was not sleepy. He thought of the power plant, even now sending its terrific stream of energy into his accumulators. He thought of the ultra radio—where could he get all the materials needed? He thought of his friends, wondering whether or not they would receive his message. He thought of Breckenridge and the other human beings who had been aboard the *Arcturus*, wondering poignantly as to their fate. He thought of Newton and of his own people, who had certainly given them up for dead long since.

But above all he thought of the beautiful, steel-true companion lying there asleep at his mailed feet, and he gazed down at her, his heart in his eyes. The firelight shone through the chinks between the boulders, casting a flickering ruddy light throughout the little cavern. Nadia lay there her head pillowed upon one strong, brown little hand. Her lips were red and sweetly curved, her cheek was smooth and firm as so much brown velvet. She was literally aglow with sheer beauty and with perfect health; and the man reflected, as he studied her hungrily, that this wild life certainly had agreed with her—she was becoming more surpassingly beautiful with every passing day.

"You little trump—you wonderful, lovely, square little brick!" he breathed silently, and bent over to touch her cheek lightly with his lips. Slight as the caress was, it disturbed her, and even in her sleep her subconscious mind sent out an exploring hand, to touch her Steve and thus be reassured. He pressed her hand and she settled back comfortably, with a long, deep breath; and he stretched his iron-clad length beside her and closed his eyes, firmly resolved not to waste a minute of this wonderful night in sleep.

When he opened them an instant later, it was broad daylight, the boulders had been rolled away, the fragrance of roasting meat permeated the atmosphere, and Nadia was making a deafening clamor, beating his steel breastplate lustily with the flat of his huge saber.

"Daylight in the swamp, you sleeper!" she exclaimed. "Roll out or roll up! Come and get it, before I throw it away!"

"I must have been kind of tired," he said sheepishly, when he saw that she had shot a bird and had cooked breakfast for them both while he had been buried in oblivion.

"Peculiar, too, isn't it?" Nadia asked, pointedly. "You only did about ten days' work yesterday in ten minutes, swinging this frightful snickersnee of yours. Why, you played with it as though it were a knitting-needle, and when I wanted to wake you up with it, I could hardly lift it."

"Thought you didn't want that subject even mentioned?" he tried to steer the talk away from his prowess with the broadsword.

"That was yesterday," airily. "Besides, I don't mind talking about you—it's thinking about us being ... you know ... that I can't stand."

"All x, ace. I get you right. Let's eat."

BREAKFAST over, they started down the valley, Stevens carrying his helmet under his arm. Hardly had they started however, than Nadia's keen eyes saw a movement through the trees, and, she stopped and pointed. Stevens looked once, then hand in hand they dashed back to their cave.

"We'll pile up some of the boulders and you lie low," he instructed her as he screwed on his helmet. She snapped open his face-plate.

"But what about you? Aren't you coming in, too?" she demanded.

"Can't—they'd surround us and starve us out. I'm safe in this armor—thank Heaven we made it as solid as we did—and I'll fight 'em in the open. I'll show 'em what the bear did to the buckwheat!"

"All right, I guess, but I wish I had my armor, too," she mourned as he snapped shut his plate and walled her into the cave with the same great rocks he had used the night before. Then, Nadia safe from attack, he drew his quiver of war-arrows into position over his shoulder, placed one at the ready on his bow-string and turned to face the horde of things rushing up the valley toward him. Wild animals he had supposed them, but as he stood firm and raised his weapon shrill whistles sounded in the throng, and he gasped as he realized that those frightful creatures must be intelligent beings, for not only did they signal to each other, but he saw that they were armed with bows and arrows, spears, and slings!

Six-limbed creatures they were, of a purplish-red color, with huge, tricornigerous heads and with staring, green, phosphorescent eyes. Two of the six limbs were always legs, two always arms; the intermediate two, due to a mid-section jointing of the six-foot-long, almost cylindrical body, could be used at will as either legs or arms. Now, out of range, as they supposed, they halted and gathered about one who was apparently their leader; some standing erect and waving four hands while shaking their horns savagely in Stevens' direction, others trotting around on four legs, busily gathering stones of suitable size for their vicious slings.

Too far away to use their own weapons and facing only one small four-limbed creature, they considered their game already in the bag, but they had no comprehension of earthly muscles, nor any understanding of the power and range of a hundred-pound bow driving a steel-headed war arrow. Thus, while they were arguing, Stevens took the offensive, and a cruelly barbed steel war-head tore completely through the body of their leader and mortally wounded the creature next beyond him. Though surprised, they were not to be frightened off, but with wild, shrill screams rushed to the attack. Stevens had no ammunition to waste, and every time that mighty bow twanged a yard-long arrow transfixed at least one of the red horde—and a body through which had torn one of those ghastly, hand-forged arrow-heads was of very little use thereafter. Accurately-sped arrows splintered harmlessly against the re-enforced windows of his helmet and against the steel guards protecting his hands. He was almost deafened by the din as the stone missiles of the slingers rebounded from his reverberating shell of steel, but he fired carefully, steadily, and powerfully until his last arrow had been loosed. Then, the wicked dirk in his left hand and the long and heavy saber weaving a circular path of brilliance in the sun, he stepped forward a couple of paces to meet the attackers. For a few moments nothing could stand before that fiercely driven blade—severed heads, limbs, and fragments of torsos literally filled the air, but sheer weight of numbers bore him down. As he fell, he saw the white shaft of one of Nadia's hunting-arrows flash past his helmet and bury itself to the flock in the body of one of the horde above him. Nadia knew that her arrows could not harm her lover, and through a chink between two boulders she was shooting into the thickest of the mob speeding her light arrows with the full power of her bow.

Though down, the savages soon discovered that Stevens was not out. In such close quarters he could not use his sword, but the fourteen-inch blade of the dirk, needle-pointed as it was and with two razor-sharp, serrated cutting edges, was itself no mean weapon, and time after time he drove it deep, taking life at every thrust. Four more red monsters threw themselves upon the prostrate man, but not sufficiently versed in armor to seek out its joints, their fierce short spear thrusts did no damage. Presently four more corpses lay still and Stevens, with his, to them incredible, earthly strength, was once more upon his feet in spite of their utmost efforts to pinion his mighty limbs, and was again swinging his devastating weapon. Half their force lying upon the field, wiped out by a small, but invincible and apparently invulnerable being, the remainder broke and ran, pursued by Stevens to the point where the red monsters had first halted. He recovered his arrows and returned to the cave, opening his face-plate as he came.

"All x, sweetheart?" he asked, rolling away the boulders. "Didn't get anything through to you, did they?"

"No, they didn't even realize that I was taking part in the battle, I guess. Did they hurt you while they had you down? I was scared to death for a minute."

"No, the old armor held. One of them must have gnawed on my ankle some, between the greave and the heel-plate, but he couldn't quite get through. 'Sa darn small opening there, too—must have bent my foot 'way around to get in at all. Have to tighten that joint up a little, I guess. I'll bet

I've got a black spot and blue spot there the size of my hand—maybe it's only the size of yours, though."

"You won't die of that, probably. Heavens, Steve, that cleaver of yours is a frightful thing in action! Suppose it's safe for us to go home?"

"Absolutely—right now is the best chance we'll ever have, and something tells me that we'd better make it snappy. They'll be back, and next time they won't be so easy to take."

"All x, then—hold me, Steve, I can't stand the sight of that—-let alone wade through it. I'm going to faint or something, sure."

"As you were!" he snapped. "You aren't going to pass out now that it's all over! It's a pretty ghastly mess, I know, but shut your eyes and I'll carry you out of sight."

"Aren't we out of sight of that place yet?" she demanded after a time.

"I have been for quite a while," he confessed, "but you're sitting pretty, aren't you? And you aren't very heavy—not here on Ganymede, anyway!"

"**P**UT me down!" she commanded. "After that crack I won't play with you any more at all—I'll pick up my marbles and go home!"

He released her and they hurried hack toward their waterfall, keeping wary eyes sharp-set for danger in any form, animal or vegetable. On the way back across the foothills Stevens shot another hexaped, and upon the plateau above the river Nadia bagged several birds and small animals, but it was not until they were actually in their own little canyon that their rapid pace slackened and their vigilance relaxed.

"After this, ace, we hunt together and we go back to wearing armor while we're hunting. It scared me out of a year's growth when you checked up missing."

"We sure do, Steve," she concurred emphatically. "I'm not going to get more than a meter away from you from now on. What do you suppose those horrible things are?"

"Which?"

"Both."

"Those flowers aren't like anything Tellus ever saw, so we have no basis of comparison. They may be a development of a flycatching plant, or they may be a link between the animal and the vegetable kingdom. However, we don't intend to study 'em, so let's forget 'em. Those animals were undoubtedly intelligent beings; they probably are a race of savages of this satellite."

"Then the really civilized races are probably...."

"Not necessarily—there may well be different types, each struggling toward civilization. They certainly are on Venus, and they once were on Mars."

"Why haven't we seen anything like that before, in all these months? Things have been so calm and peaceful that we thought we had the whole world to ourselves, as far as danger or men were concerned."

"We never saw them before because we never went where they lived—you were a long ways from your usual stamping-grounds, you know. That animal-vegetable flower is probably a high-altitude organism, living in the mountains and never coming as low as we are down here. As for the savages—whatever they are—they probably never come within five kilometers of the falls. Many primitive peoples think that waterfalls are inhabited by demons, and maybe these folks are afflicted the same way."

"We don't know much about our new world yet, do we?"

"We sure don't—and I'm not particularly keen on finding out much more about it until we get organized for trouble, either. Well, here we are—just like getting back home to see the 'Hope,' isn't it?"

"It *is* home, and will be until we get one of our own on earth," and after Stevens had read his meters, learning with satisfaction that the full current was still flowing into the accumulators, he began to cut up the meat.

"Now that you've got the power-plant running at last, what next?" asked Nadia, piling the cuts in the freezer.

"Brandon's ultra-radio comes next, but it's got more angles to it than a cubist's picture of a set of prisms; so many that I don't know where to begin. There, that job's done—let's sit down and I'll talk at you awhile. Maybe between us we can figure out where to start. I've got everything to build it lined up except for the tube, but that's got me stopped cold. You see, fields of force are all right in most places, but I've got to have one tube, and it's got to have the hardest possible vacuum. That means a mercury-vapor super pump. Mercury is absolutely the only thing that will do the trick and the mercury is one thing that is conspicuous by its absence in these parts. So are tungsten for filaments, tantalum for plates, and platinum for leads; and I haven't found anything that I can use as a getter, either—a metal, you know, to flash inside the tube to clean up the last traces of atmosphere in it."

"I didn't suppose that such a simple thing as a radio tube could hold you up, after the perfectly unbelievable things that you have done already—but I see now how it could. Of course, the tubes in our receiver over there are too small?"

"Yes, they are only receiver and communicator tubes, and I need a high-power transmitting tube—a fifty-kilowatter, at least. I'd give my left leg to the knee joint for one of those big water-cooled, sixty-kilowatt ten-nineteens right now—it would save us a lot of grief."

"Maybe you could break up those tubes and use the plates and so on?"

"I thought of that, but it won't work—there isn't half enough metal in the lot, and the filaments in particular are so tiny that I couldn't possibly work them over into a big one. Then, too, we haven't got many spare tubes, and if I smash the ones we're using, I put our communicators out of business for good, so that we can't yell for help if we have to drift home—and I still don't get any mercury."

"Do you mean to tell me there's no mercury on this whole planet?"

"Not exactly; but I do mean that I haven't been able to find any, and that it's probably darned scarce. And since all the other metals I want worst are also very dense and of high atomic weight, they're probably mighty scarce here, too. Why? Because we're on a satellite, and no matter what hypothesis you accept for the origin of satellites, you come to the same conclusion—that heavy metals are either absent or most awfully scarce and buried deep down toward the center. There are lots of heavy metals in Jupiter somewhere, but we probably couldn't find them. Jupiter's atmosphere is one mass of fog, and we couldn't see, since we haven't got an infra-red transformer. I could build one, in time, but it would take quite a while—and we couldn't work on Jupiter, anyway, because of its gravity and probably because of its atmosphere. And even if we could work there, we don't want to spend the rest of our lives prospecting for mercury." Stevens fell silent, brow wrinkled in thought.

"You mean, dear, that we're..." Nadia broke off, the sentence unfinished.

"Gosh, no! There's lots of things not tried yet, and we can always set out to drift it. I was thinking only of building the tube. And I'm trying to think ... say, Nadia, what do you know about Cantrell's Comet?"

"Not a thing, except that I remember reading in the newspapers that it was peculiar for something or other. But what has Cantrell's Comet got to do with the high cost of living—or with radio tubes? Have you gone cuckoo all of a sudden?"

"You'll be surprised!" Stevens grinned at her puzzled expression. "Cantrell's Comet is one of Jupiter's comet family and is peculiar in being the most massive one known to science. It was hardly known until after they built those thousand-foot reflectors on the Moon, where the seeing is always perfect, but it has been studied a lot since then. Its nucleus is small, but extremely heavy—it seems to have an average density of somewhere around sixteen. There's platinum and everything else that's heavy there, girl! They ought to be there in such quantity that even such a volunteer chemist as I am could find them!"

"HEAVENS, Steve!" A look of alarm flashed over Nadia's face, then disappeared as rapidly as it had come into being. "But of course, comets aren't really dangerous."

"Sure not. A comet's tail, which so many people are afraid of as being poison gas, is almost a perfect vacuum, even at its thickest, and we'd have to wear space-suits anyway. And speaking of vacuum ... whoopee! We don't need mercury any more than a goldfish needs a gas-mask. When we get Mr. Tube done, we'll take him out into space, leaving his mouth open, and very shortly he'll be as empty as a flapper's skull. Then we'll seal him up, flash him out, come back here, and start spilling our troubles into Brandon's shell-like ear!"

"Wonderful, Steve! You do get an idea occasionally, don't you? But how do we get out there? Where is this Cantrell's Comet?"

"I don't know, exactly—there's one rub. Another is that I haven't even started the transmitter and receptor units. But we've got some field-generators here on board that I can use, so it won't be so bad. And our comet is in this part of the solar system somewhere fairly close. Wish we had an Ephemeris, a couple of I-P solar charts, and a real telescope."

"You can't do much without an Ephemeris, I should think. It's a good thing you kept the chronometers going. You know the I-P time, day, and dates, anyway."

"I'll have to do without some things, that's all," and the man stared absently at the steel wall. "I remember something about its orbit, since it is one thing that all I-P vessels have to steer clear of. Think I can figure it close enough so that we'll be able to find it in our little telescope, or even on our plate, since we'll be out of this atmosphere. And it might not be a bad idea for us to get away, anyway. I'm afraid of those folks on that space-ship, whoever they were, and they must live around here somewhere. Cantrell's Comet swings about fifty million kilometers outside Jupiter's orbit at aphelion—close enough for us to reach, and yet probably too far for them to find us easily. By the time we get back here, they probably will have quit looking for us, if they look at all. Then too, I expect these savages to follow us up. What say, little ace—do we try it or do we stay here?"

"You know best, Steve. As I said before, I'm with you from now on, in whatever you think best to do. I know that you think it best to go out there. Therefore, so do I."

"Well," he said, finally, "I'd better get busy, then—there's a lot to do before we can start. The radio doesn't come next, after all—the transmitter and receptor units come ahead of it. They won't mean wasted labor, in any event, since we'll have to have them in case the radio fails. You'd better lay in a lot of supplies while I'm working on that stuff, but don't go out of sight, and yell like fury if you see anything. We'd both better wear full armor every time we go out-of-doors—unless I'm all out of control we aren't done with those savages yet. Even though they may be afraid of the demons of the falls, I think they'll have at least one more try at us."

While Nadia brought in meat and vegetables and stored them away, Stevens attacked the problem of constructing the pair of tight-beam, auto-dirigible transmitter and receptor units which would connect his great turbo-alternator to the accumulators of their craft, wherever it might be in space. From the force-field generators of the "Forlorn Hope" he selected the two most suitable for his purpose, tuned them to the exact frequency he required, and around them built a complex system of condensers and coils.

Day after day passed. Their larder was full, the receptor was finished, and the beam transmitter was almost ready to attach to the turbo-alternator before the calm was broken.

"Steve!" Nadia shrieked. Glancing idly into the communicator plate, she had been perfunctorily surveying the surrounding territory. "They're coming! Thousands of them! They're all over the bench up there, and just simply pouring down the hills and up the valley!"

"Wish they'd waited a few hours longer—we'd have been gone. However, we're just about ready for them," he commented grimly, as he stared over her shoulder into the communicator plate. "We'll make a lot of those Indians wish that they had stayed at home with their papooses."

"Have you got all those rays and things fixed up?"

"Not as many as I'd like to have. You see, I don't know the composition of the I-P ray, since it is outlawed to everybody except the police. Of course I could have found out from Brandon, but never paid any attention to it. I've got some nice ultra-violet, though, and a short-wave oscillatory that'll cook an elephant to a cinder in about eight seconds. We'll keep them amused, no fooling! Glad we had time to cover our open sides, and it looks as though that meteorite armor we put over the projectors may be mighty useful, too."

On and on the savages came, massed in formations showing some signs of rude discipline. This time there was neither shrieking nor yelling; the weird creatures advanced silently and methodically. Here and there were massed groups of hundreds, dragging behind them engines which Stevens studied with interest.

"Hm ... m ... m. Catapults," he mused. "You were right, girl of my dreams—armor and bows and arrows wouldn't help us much right now. They're going to throw rocks at us that'll have both mass and momentum. With those things they can cave in our side-armor, and might even dent our roof. When one of those projectiles hits, we want to know where it ain't, that's all."

Stevens cast off the heavily-insulated plug connecting the power plant leads to his now almost fully charged accumulators, strapped himself and Nadia into place at the controls, and waited, staring into the plate. Catapult after catapult was dragged to the lip of the little canyon, until six of them bore upon the target. The huge stranded springs of hair, fiber, and sinew were wound up to the limit, and enormous masses of rock were toilsomely rolled upon the platforms. Each "gunner" seized his trip, and as the leader shrieked his signal the six ponderous masses of metalliferous rock heaved into the air as one. But they did not strike their objective, for as the signal was given, Stevens shot power into his projectors. The "Forlorn Hope" leaped out of the canyon and high into the air over the open meadow, just as the six great projectiles crashed into the ground upon the spot which, an instant before, she had occupied.

RUDIMENTARY discipline forgotten, the horde rushed down into the canyon and the valley, in full clamor of their barbaric urgings. Horns and arms tossed fiercely, savage noises rent the air, and arrows splintered harmlessly upon steel plate as the mystified and maddened warriors upon the plain below gave vent to their outraged feelings.

"Look, Nadia! A whole gang of them are smelling around that power plug. Pretty soon somebody's going to touch a hot spot, and when he does, we'll cut loose on the rest of them."

The huge insulating plug, housing the ends of the three great cables leading to the converters of the turbo-alternator, lay innocently upon the ground, its three yawning holes invitingly open to savage arms. The chief, who had been inspecting the power-plant, walked along the triplex lead and joined his followers at its terminus. Pointing with his horns, he jabbered orders and three red monsters, one at each cable, bent to lift the plug, while the leader himself thrust an arm into each of the three contact holes. There was a flash of searing flame and the reeking smoke of burning flesh—those three arms had taken the terrific no-load voltage of the three-phase converter system, and the full power of the alternator had been shorted directly to ground through the comparatively small resistance of his body.

Stevens had poised the "Forlorn Hope" edgewise in mid-air, so that the gleaming, heavily armored parabolic reflectors of his projectors, mounted upon the leading edge of the fortress, covered the scene below. As the charred corpse of the savage chieftain dropped to the ground, it seemed to the six-limbed creatures that the demons of the falls had indeed been annoyed beyond endurance by their intrusion; for, as if in response to the flash of fire from the power plug, that structure so peculiarly and so stolidly hanging in the air came plunging down toward them. From it there reached down twin fans of death and destruction: one flaming and almost invisibly incandescent violet which tore at the eyes and excruciatingly disintegrated brain and nervous tissues; the other dully glowing an equally invisible red, at the touch of which body temperature soared to lethal heights and foliage burst cracklingly into spontaneous flame.

In their massed hundreds, the savages dropped where they stood, life rived away by the torturing ultra-violet, burned away by the blast of pure heat, or consumed by the conflagrations that raged instantly wherever that wide-sweeping fan encountered combustible material. In the face of power supernatural they lost all thought of attack or of conquest, and sought only and madly to escape. Weapons were thrown away, the catapults were abandoned, and, every man for himself, the mob fled in wildest disorder, each striving to put as much distance as possible between himself and that place of dread mystery, the waterfall.

"Well, I guess that'll hold 'em for a while," Stevens dropped their craft back into its original quarters in the canyon. "Whether they ever believed before that this falls was inhabited by devils or not, they think so now. I'll bet that it will be six hundred Jovian years before any of them ever come within a hundred kilometers of it again. I'm glad of it, too, because they'll let our power plant alone now. Well, let's get going—we've got to make things hum for a while!"

"Why all the rush? You just said that we have scared them away for good."

"The savages, yes, but not those others. We've just turned loose enough radiation to affect detectors all over the system, and it's up to us to get this beam projector set up, get away from here, and get our power shut off before they can trace us. Snap it up, ace!"

The transmitter unit was installed at the converters, the cable was torn out, and, having broken the last material link between it and Ganymede, Stevens hurled the "Forlorn Hope" out into space, using the highest acceleration Nadia could endure. Hour after hour the massive wedge of steel bored outward, away from Jupiter; hour after hour Stevens' anxious eyes scanned his instruments; hour after hour hope mounted and relief took the place of anxiety as the screens remained blank throughout every inquiring thrust into the empty ether. But they knew they would have to keep sharp vigilance.

Continuing a Thrilling New Serial of Interplanetary Life and Travel by Edward E. Smith, Ph.D.

Author of "Skylark of Space," and "Skylark Three"

PART II

O NE of the most fascinating mysteries of the heavens is the comet. It goes through space, gets near enough to the earth to be seen, and then goes off and disappears in celestial distance. Often it has a hyperbolic orbit, which would make it impossible to come back. Yet it may return – apparently contradicting the geometry of conic sections. This only goes to prove once more that it is risky to say anything is impossible – even that our hero of this story manages beautifully, with the aid of Cantrell's Comet, to avoid complete annihilation while stranded in interstellar space.

Read "what went before" and then continue the second instalment.

What Went Before:

THE Interplanetary Vessel Arcturus sets out for Mars, with Breckenridge as chief pilot, carrying on board, besides its regular crew and some passengers, the famous Dr. Stevens, designer of space ships and computer. He checks computations made by astronomers stationed in floating observatories, and after he has located any trouble and suggests a plan for minimizing the hazards of the trip from the earth to Mars, he reports his findings and suggestions to Mr. Newton, chief of the Interplanetary Corporation.

Stevens then takes Nadia, Mr. Newton's beautiful young daughter, on a specially conducted sight-seeing tour of the Arcturus and thoroughly explains to her all of the works of the vessel. Nadia has herself had a good science education. While they are down at the bottom of the ship—nearing the end of their tour—Stevens feels a barely perceptible movement of the vessel from its course. When he turns on the visiplate, he is horrified to find that a mysterious ray of unparalleled power has neatly sliced the Arcturus in several places.

Nadia and Stevens are completely separated from the rest of the crew and passengers of the ship, so they get into a lifeboat, which is equipped for a limited amount of space travel. Despite the strict and apparently effective vigilance of the enemy destroyer, Stevens and Nadia make their getaway in the lifeboat, which they aptly call "Forlorn Hope," and finally make a safe landing on Ganymede, where Stevens plans to build a power-plant and a radio transmitter, to enable him to communicate with the earth or with the IPV Sirius, which is used by Westfall and Brandon (two of the world's best scientists) as a floating laboratory.

With the very scant apparatus and material available, Stevens sets to work on his power plant. Just as they have it completed and ready to start for Cantrell's Comet, where Stevens believes he can obtain the necessary metal for his giant transmitting tube, they experience a close call with carnivorous plants on the satellite and later with savage inhabitants, which precipitates their trip to the comet.

CHAPTER V
Cantrell's Comet

FAR out in space, Jupiter, a tiny moon and its satellites mere pin-points of light, Stevens turned to his companion with a grin.

"Well, Nadia, old golf-shootist, here's where we turn spacehounds again. Hope you like it better this time, because I'm afraid that we'll have to stay weightless for quite a while." He slowly throttled down the mighty flow of power, and watched the conflicting emotions play over Nadia's face in her purely personal battle against the sickening sensations caused by the decrease in their acceleration.

"I'm sorry as the dickens, sweetheart," he went on, tenderly, and the grin disappeared. "Wish I could take it for you, but...."

"But there are times when we've got to fight our own battles and bury our own dead," she interrupted, gamely. "Cut off the rest of that power! I'm *not* going to be sick—I *won't* be a—what do you spacehounds call us poor earth-bound dubs who can't stand weightlessness—weight-fiends, isn't it?"

"Yes; but you aren't...."

"I know I'm not, and I'm not going to be one, either! I'm all x, Steve—it's not so bad now, really. I held myself together that time, anyway, and I feel lots better now. Have you found Cantrell's Comet yet? And why so sure all of a sudden that they can't find us? That power beam still connects us to Ganymede, doesn't it? Maybe they can trace it."

"At-a-girl, ace!" he cheered. "I'll tell the world you're no weight-fiend—you're a spacehound right. Most first-trippers, at this stage of the game, wouldn't be caring a whoop whether school kept or not, and here you're taking an interest in all kinds of things already. You'll do, girl of my heart—no fooling!"

"Maybe, and maybe you're trying to kid somebody," she returned, eyeing him intently. "Or maybe you just don't want to answer those questions I asked you a minute ago."

At the bottom of a shaft a section of the rocky wall swung aside, revealing the yawning black mouth of a horizontal tunnel. At intervals upon its roof there winked into being almost invisible points of light. Along that line of lights the life-boats felt their way, coming finally into a huge cavern....

"No, that's straight data, right on zero across the panel," he assured her. "And as for your questions, they're easy. No, I haven't looked for the comet yet, because we'll have to drift for a couple of days before we'll be anywhere near where I think it is. No, they can't trace us, because there is now nothing to trace, unless they can detect the slight power we are using in our lights and so on—which possibility is vanishingly small. Potentially, our beam still exists, but since we are drawing no power, it has no actual present existence. See?"

"Uh-uh," she dissented. "I can't say that I can quite understand how a beam can exist potentially and yet not be there actually enough to trace. Why, a thing has to be actual or not exist at all—you can't possibly have something that is nothing. It doesn't make sense. But lay off those integrations of yours, please," as now armed with a slate-pencil, Stevens began to draw a diagram upon a four-foot sheet of smooth slate. "You know that your brand of math is over my head like a circus tent, so we'll let it lie. I'll take your word for it. Steve—if you're satisfied, it's all x with me."

"I think I can straighten you out a little, by analogy. Here's a rough sketch of a cylinder, with shade and shadow. You've had descriptive geometry, of course, and so know that a shadow, being simply a projection of a material object upon a plane, is a two-dimensional thing—or rather, a two-dimensional concept. Now take the shade, which is, of course, this entire figure here, between the cylinder casting the shadow and the plane of projection. You simply imagine that there is a point source of light at your point of projection: it isn't really there. The shade, then, of which I am drawing a picture, has only a potential existence. You know exactly where it is, you can draw it, you can define it, compute it, and work with it—but still it doesn't exist; there is absolutely nothing to differentiate it from any other volume of air, and it cannot be detected by any physical or mechanical means. If, however, you place a light at the point of projection, the shade becomes actual and can be detected optically. By a sufficient stretch of the imagination, you might compare our beam to that shade. When we turn our power on, the beam is actual; it is a stream of tangible force, and as such can be detected electrically. When our switches here are open, however, it exists only potentially. There is no motion in the ether, nothing whatever to indicate that a beam had ever actually existed there. With me?"

"Floundering pretty badly, but I see it after a fashion. You physicists are peculiar freaks—where we ordinary mortals see actual, solid, heavy objects, you see only empty space with a few electrons and things floating around in it; and yet where we see only empty space, you can see things 'potentially' that may never exist at all. You'll be the death of me yet, Steve! But I'm wasting a lot of time. What do we do now?"

"We get busy on the big tube. You might warm up the annealing oven and melt me that pot of glass, while I get busy on the filament supports, plate brackets, and so on." Both fell to work with a will, and hours passed rapidly and almost silently, so intent was each upon his own tasks.

"All x, Steve." Nadia broke the long silence. "The pyrometer's on the red, and the oven's hot," and the man left his bench. Taking up a long paddle and an even longer blowpipe, he skimmed the melt to a dazzlingly bright surface and deftly formed a bubble.

"I just love to talk at you when you've got your mouth full of a blowpipe." Nadia eyed him impishly and tucked her feet beneath her, poised weightless as she was. "I've got you foul now—I can say anything I want to, and you can't talk back, because your bubble will lose its shape if you do. Oh, isn't that a beauty! I never saw you blow anything that big before," and she fell silent, watching intently.

Slowly there was being drawn from the pot a huge, tapering bulb of hot, glistening glass, its cross-section at the molten surface varying as Stevens changed the rate of draw or the volume of air blown through the pipe. Soon that section narrowed sharply. The glass-blower waved his hand and Nadia severed the form neatly with a glowing wire, just above the fluid surface of the glass remaining in the pot. Pendant from the blowpipe, the bulb was placed over the hot-bench, where Stevens, now begoggled, begloved, and armed with a welding torch, proceeded to fuse into the still, almost plastic, glass sundry necks, side-tubes, supports and other attachments of peculiar pattern. Finally the partially assembled tube was placed in the annealing oven, where it would remain at a high and constant temperature until its filaments, grids, and plates had been installed. Eventually, in that same oven, it would be allowed to cool slowly and uniformly over a period of days.

THUS were performed many other tasks which are ordinarily done either by automatic machinery or by highly skilled specialists in labor—for these two, thrown upon their own resources, had long since learned how much specialization may be represented by the most commonplace article. Whenever they needed a thing they did not have—which happened every day—they had either to make it or else, failing in that, to go back and build something that would enable them to manufacture the required item. Such setbacks had become so numerous as to be expected as part of the day's work; they no longer caused exasperation or annoyance. For two days the two jacks-of-all-trades worked at many lines and with many materials before Stevens called a halt.

"All x, Nadia. It's time for us to stop tinkering and turn into astronomers. We've been out for fifty I-P hours, and we'd better begin looking around for our heap of scrap metal," and, the girl at the communicator plate and Stevens at their one small telescope, they began to search the black, star-jeweled heavens for Cantrell's Comet.

"According to my figures, it ought to be about four hours right ascension, and something like plus twenty degrees declination. My figures aren't accurate, though, since I'm working purely from memory, so we'd better cover everything from Aldebaran to the Pleiades."

"But the directions will change as we go along, won't they?"

"Not unless we pass it, because we're heading pretty nearly straight at it, I think."

"I don't see anything interesting thereabouts except stars. Will it have much tail?"

"Very little—it's close to aphelion, you know, and a comet doesn't have much of a tail so far away from the sun. Hope it's got some of its tail left, though, or we may miss it entirely."

Hours passed, during which the two observers peered intently into their instruments, then Stevens left the telescope and went over to his slate.

"Looks bad, ace—we should have spotted it before this. Time to eat, too. You'd better...."

"Oh, look here, quick!" Nadia interrupted. "Here's something! Yes, it *is* a comet, and quite close—it's got a little bit of a dim tail."

Stevens leaped to the communicator plate, and, blond head pressed close to brown, the two wayfarers studied the faint image of the wanderer of the void.

"That's it, I just *know* it is!" Nadia declared. "Steve, as a computer, you're a blinding flash and a deafening report!"

"Yeah—missed it only about half a million kilometers or so," he replied, grinning, "and I'd fire a whole flock of I-P check stations for being four thousand off. However, I could have done worse—I could easily have forgotten all the data on it, instead of only half of it." He applied a normal negative acceleration, and Nadia heaved a profound sigh of relief as her weight returned to her and her body again became manageable by the ordinary automatic and involuntary muscles.

"Guess I am a kind of a weight-fiend at that, Steve—this is much better!" she exclaimed.

"Nobody denies that weight is more convenient at times; but you're a spacehound just the same—you'll like it after a while," he prophesied.

Stevens took careful observations upon the celestial body, altered his course sharply, then, after a measured time interval, again made careful readings.

"That's it, all x," he announced, after completing his calculations, and he reduced their negative acceleration by a third. "There—we'll be just about traveling with it when we get there," he said. "Now, little K. P. of my bosom, our supper's been on minus time for hours. What say we shake it up?"

"I check you to nineteen decimals," and the two were soon attacking the savory Ganymedean goulash which Nadia had put in the cooker many hours before.

"Should we both go to sleep, Steve, or should one of us watch it?"

"Sleep, by all means. There's no meteoric stuff out here, and we won't arrive before ten o'clock tomorrow, I-P time," and, tired out by the events of the long day, man and maid sought their beds and plunged into dreamless slumber.

While they slept, the "Forlorn Hope" drove on through the void at a terrific but constantly decreasing velocity; and far off to one side, plunging along a line making a sharp angle with their own course, there loomed larger and larger the masses which made up the nucleus of Cantrell's Comet.

Upon awakening, Stevens' first thought was for the comet, and he observed it carefully before he aroused Nadia, who hurried into the control room. Looming large in the shortened range of the plate, their objective hurtled onward in its eternal course, its enormous velocity betrayed only by the rapidity with which it sped past the incredibly brilliant background of infinitely distant stars. Apparently it was a wild jumble of separate fragments; a conglomerate, heterogeneous aggregation of rough and jagged masses varying in size from grains of sand up to enormous chunks, which upon Earth would have weighed millions of tons. Pervading the whole nucleus, a slow, indefinite movement was perceptible—a vague writhing and creeping of individual components working and slipping past and around each other as they all rushed forward in obedience to the immutable cosmic law of gravitation.

"Oh, isn't that wonderful!" Nadia breathed. "Think of actually going to visit a comet! It sort of scares me, Steve—it's so creepy and crawly looking. We're awfully close, aren't we?"

"Not so very. We'd probably have lots of time to eat breakfast. But just to be on the safe side, maybe I'd better camp here at the board, and you bring me over something to eat."

"All x, Chief!" and Stevens ate, one eye upon the screen, watching closely the ever-increasing bulk of the comet.

FOR many minutes he swung the *Forlorn Hope* in a wide curve approaching the mountain of metal ever and ever more nearly, then turned to the girl.

"Hold everything, Nadia—power's going off in a minute!" He shut off the beam; then, noting that they were traveling a trifle faster than the comet, he applied a small voltage to one dirigible projector. Darting the beam here and there, he so corrected their flight that they were precisely stationary in relation to the comet. He then opened his switches, and the *Forlorn Hope* hurtled on. Apparently motionless, it was now a part of Cantrell's Comet, traveling in a stupendous, elongated ellipse about the Master of our Solar System, the Sun.

"There, ace, who said anything about weight-fiends? I was watching you, and you never turned a hair that time."

"Why, that's right—I never even thought about it—I was so busy studying that thing out there! I suppose I've got used to it already?"

"Sure—you're one of us now. I knew you would be. Well, let's go places and do things! You'd better put on a suit, too, so you can stand in the air-lock and handle the line."

They donned the heavily insulated, heated suits, and Stevens snapped the locking plugs of the drag line into their sockets upon the helmets.

"Hear me?" he asked. "Sound-disks all x?"

"All x."

"On the radio—all x?"

"All x."

"I tested your tanks and heaters—they're all x. But you'll have to test...."

"I know the ritual by heart, Steve. It's been in every show in the country for the last year, but I didn't know you had to go through it every time you went out-of-doors! Halves, number one all x, two all x, three all x...."

"Quit it!" he snapped. "You aren't testing those valves! That check-up is no joke, guy. These suits are complicated affairs, and some parts are apt to get out of order. You see, a thing to give you fresh air at normal pressure and to keep you warm in absolute space can't be either simple or fool-proof. They've worked on them for years, but they're pretty crude yet. They're tricky, and if one goes sour on you, out in space, it's just too bad—you're lucky to get back alive. A lot of men are still out there somewhere because of the sloppy check-ups."

"'Scuse it, please—I'll be good," and the careful checking and testing of every vital part of the space-suits went on.

Satisfied at last that the armor was spaceworthy, Stevens picked up the coils of drag-line, built of a non-metallic fiber which could retain its flexibility and strength in the bitter cold of outer space, and led the girl into the air-lock.

"Heavens, Steve! It's perfectly stupendous, and grinding around worse than the wreckage of the *Arcturus* was when I wouldn't let you climb up it—why, I thought comets were *little*, and hardly massive at all!" exclaimed the girl.

"This is little, compared to any regular planet or satellite or even to the asteroids. There's only a few cubic kilometers of matter there, and, as I said before, it's a decidedly unusual comet. You know the game?"

"I've got it—and believe me, I'll yank you back here a lot faster than you can jump over there if any one of those lumps starts to fall on you! Is this drag line long enough?"

"Yes, I've got a hundred meters here, and it's only fifty meters over there to where I'm going. So long," and with a light thrust of his feet, he dove head foremost across the intervening space, a heavy pike held out ahead of him. Straight as a bullet he floated toward his objective, a jagged chunk many yards in diameter, taking the shock of his landing by sliding along the pike-handle as its head struck the mass.

Then, bracing his feet against one lump, he pushed against its neighbor, and under that steady pressure the enormous masses moved apart and kept on moving, grinding among their fellows. Over and around them Stevens sprang, always watching his line of retreat as well as that of his advance, until his exploring pike struck a lump of apparently solid metal. Hooking the fragment toward him, he thrust savagely with his weapon and was reassured—that object was not only metal, but it was metal so hard that his pike-head of space-tempered alloy steel did not make an impression upon its surface. Turning on his helmet light he swung his heavy hammer repeatedly but could not break off even a small fragment.

"Found something, Steve?" Nadia's voice came clearly in his ears.

"I'll say I have! A hunk of solid, non-magnetic metal about the size of an office desk. I can't break off any of it, so I guess we'll have to grab the whole chunk."

He hitched the end of his cable around the nugget, made sure that the loops would not slip, and then, as Nadia tightened the line, he shoved mightily.

"All x, Nadia, she's coming! Pull in my drag line as I said over there, and I'll help you land her."

Inside the *Forlorn Hope* the mass of metal was urged into the shop, where Stevens clamped it immovably to the steel floor, before he took off his space-suit.

"Why, it's getting covered with snow, and the whole room is getting positively *cold*!" Nadia exclaimed.

"Sure. Anything that comes in from space is cold, even if it's been out only a few minutes, and that hunk of stuff has been out for nobody knows how many million years. It didn't get much heat from the sun except at perihelion, you know, so it's probably somewhere around minus two hundred and sixty degrees now. I'll have to throw a heater on it for half an hour before we can touch it. And since this is more or less new stuff to you, I'll caution you—don't try to touch anything that has just come in. That hammer or pike would freeze your hand instantly, even though they've been out only a little while. Before you touch anything, blow on it, like this, see? If your breath freezes solid on it, like that, don't touch it—it's cold."

UNDER the infra-beams of the heater, the mass of the metal was brought to room temperature and Stevens attacked it with his machine tools. Bit by bit the stubborn material was torn from the lump. Through heavy goggles he watched the incandescent mass in a refractory crucible, in the heart of the induction furnace.

"What do you think you've got—what you want?"

"I don't know. It wasn't iron—it wouldn't hold a magnet. It's royal metal of some kind, I think. Base metals mostly melt at around fifteen hundred, and that crucible is still dry as a bone at better than seventeen."

"How are you going to separate out the tantalum and the others you want from the ones that you don't want?"

"I'm afraid that I'm not going to, very well," replied Stevens, with a wry grimace. "What I don't know about metallurgy would fill a library, and I'm probably the world's worst chemist. However, by a series of successive liquations, I hope to separate out fractions that I can use. Platinum melts somewhere around seventeen-fifty, tantalum about twenty-nine hundred, and tungsten not until 'way up around thirty-three, or four hundred—and that, by the way, means lots of grief. Of course, each fraction will probably be an alloy of one kind or another, but I think maybe I'll be able to make them do."

"But mayn't that whole chunk be a pure metal?"

"It's conceivable, but not probable. There, she's beginning to separate at just below eighteen hundred! Platinum group coming out now, I think—platinum, rhodium, iridium, and that gang, you know. While I'm doing this, you might be getting those five coils into exact resonance, if you want to."

"Sure I want to," and Nadia made her way across to the short-wave oscillator and set to work.

After an hour or so, bent over her delicate task, she began to twitch uneasily, then shrugged her shoulders impatiently.

"What's the idea of staring at me so?" she broke out suddenly. "How do you expect me to tune these things up if you...." She stopped abruptly, mouth open in amazement, as she turned toward Stevens. He had not been looking at her, but he turned a surprised face from his own task at the sound of her voice. "Excuse me, please, Steve. I don't know what's the matter with me—must be getting jumpy, I guess."

"I wish that was all, but it isn't!" Face suddenly grim and hard, Stevens leaped to the communicator plate and shot the beam out into space. "There's an answer, but that isn't it. You're a fine-tuned instrument yourself, ace, and you've detected something.... I thought so! There's the answer—the guy that was looking at you!"

Plainly there was revealed upon the plate a small, spherical space-ship, very like the one that had attacked and destroyed the *Arcturus*. After Nadia had taken one glance at it, Stevens shut off the power and leaped out into the shop. He closed all the bulkhead doors and air-break openings, then closed and secured the massive insulating door of the lifeboat in which they had made their headquarters. Then, after they had again put on the space-suits they had taken off such a short time before, he extinguished all the lights and hooded the communicator screen before he ventured again to glance out into the void.

"If I had a brain in my head, instead of the pint of bean soup I've got up there, we'd have worn these when they cut up the *Arcturus*, and saved us a lot of mental wear and tear," he remarked. "They were right there in the lockers all the time, and I knew it!"

"Well, we got away, anyway. You couldn't be expected to think of everything at once. We didn't have much time, you know."

"No, but I should have thought of anything as obvious as that, anyway. Wonder how they found us? Did they detect us, or did they come out to this comet after metal, same as we did, and find us accidentally? However, it all works out the same—they're apparently out to get us. I'm afraid this is going to be a whole lot like a rabbit fighting back at a man with a gun; but we'll sure try to nibble us off a lunch while they're getting a square meal ... here they come!"

The enemy sphere launched its flaming plane of force, and the *Forlorn Hope* shuddered in every plate and member as its apex was severed cleanly under the impact. Instantly Stevens hurled his only weapons. Flaming ultra-violet and dully glowing infra-red, the twin beams lashed out; but their utmost force was of slight moment to the enormous power driving the enemy screens. Two circular spots of cherry red in space were the only results of Stevens' attack, and the next fierce cut sheared away the two projectors and, incidentally, a full half of the fifty-inch armor of the leading edge.

"Then we're checking out now?" Nadia asked quietly, as the man's hands dropped from his useless controls. "I'm sorrier than I can say, lover. But at least, I'm glad that I can go out with you," and her glorious eyes were shining with unshed tears.

"Maybe, but snap out of it, girl—our hearts are still beating! We're not dead yet, and maybe we won't be. Perhaps they want to capture us alive, as they did before; if so, we may be able to hide out on them somewhere and pull off another escape. Things don't look very bright, I know, but we're not checking out until our numbers are actually run up!"

He hooked a hand under her belt as the shocks came closer, and stood tense and ready. The lancing plane cut through one end of their control room, and Stevens leaped with his companion toward the new-made opening; while the air shrieked outward into space and their suits bulged suddenly with the abrupt increase in pressure differential. While they were in midflight, the frightful blade of destruction cleaved its way through the control board and through the spot upon which they had been standing a moment before. As they passed the severed edge, en route into open spare, Stevens seized a metal brace and clung there, every nerve taut.

"Something funny here, Nadia," he said after a little, in a low tone. "They should have made one more cut, to make us absolutely blind and helpless. As it is, they've clipped off all our projectors, so we can't move, but I think we've got the whole control compartment of number two lifeboat untouched. If so, we can look around, anyway. Let's go!"

Floating without effort from fragment to fragment, they made their way toward the section of their cruiser as yet undamaged. They found an airlock in working order, and were soon in the second lifeboat, where Stevens hastily turned on a communicator and peered out into space.

"There they are! There's another stranger out there, too. They're fighting with her, now—that's probably why they didn't polish us off." Steel-braced, clumsy helmets touching, the two Terrestrials stared spell-bound into the plate; watching while the insensately vicious intelligences within the sphere brought its every force to bear upon another and larger sphere which was now so close as to be plainly visible. Like a gigantic drop of quicksilver this second globe appeared—its smooth and highly-polished surface one enormous, perfect, spherical mirror. Watching tensely, they saw flash out that frightful plane of seething energy, with the effects of which they were all too familiar, and saw it strike full upon the dazzling ball.

"This is awful, ace!" Stevens groaned. "They haven't got ray-screens, either, and without them they don't stand a chance. No possible substance can stand up under that beam. When they get done and turn back to us, we'll have to dive back to where we were."

BUT that brilliant mirror was not as vulnerable as Stevens had supposed. The plane of force struck and clung, but could not penetrate it. Broken up into myriads of scintillating crystals of light, intersecting, multi-colored rays, and cascading flares of sparkling energy, the beam was reflected, thrown back, hurled away on all sides into space in coruscating, blinding torrents. And neither was the monster globe inoffensive. The straining watchers saw a port open suddenly, emit a flame-erupting something, and close as rapidly as it had opened. That something was a projectile, its propelling rockets fiercely aflame; as smoothly brilliant as its mother-ship and seemingly as impervious to the lethal beams of the common foe. Detected almost instantly as it was, it received the full power of the savage attack. The hitherto irresistible plane of force beat upon it; ultra-violet, infra-red, and heat rays enveloped it; there were hurled against it all the forces known to the scientific minds within that fiendishly destructive sphere.

Finally, only a scant few hundreds of yards from its goal, the protective mirror was punctured and the freight of high explosive let go, with a silent, but nevertheless terrific, detonation. But now another torpedo was on its way, and another, and another; boring on ruthlessly toward the smaller sphere. Fighting simultaneously three torpedos and the giant globe, the enemy began dodging, darting hither and thither with a stupendous acceleration; but the tiny pursuers could not be shaken off. At every dodge and turn, steering rockets burst into furious activity and the projectiles rushed ever nearer. Knowing that she had at last encountered a superior force, the sphere turned in mad flight; but, prodigious as was her acceleration, the torpedoes were faster and all three of them struck her at once. There ensued an explosion veritably space-racking in its intensity; a flash of incandescent brilliance that seemed to fill all space, subsiding into a vast volume of tenuous gas which, feebly glowing, flowed about and attached itself to Cantrell's Comet. And in the space where had been the enemy sphere, there was nothing.

A slow-creeping pale blue rod of tangible force reached out from the great sphere, touched the wreckage of the *Forlorn Hope*, and pulled; gently, but with enormous power.

"Tractor beams again!" exclaimed Stevens, still at the plate. "Everybody's got 'em but us, it seems."

"And we can't fight a bit any more, can we?"

"Not a chance—bows and arrows wouldn't do us much good. However, we may not need 'em. Since they fought that other crew, and haven't blown us up, they aren't active enemies of ours, and may be friendly. I haven't any idea who or what they are, since even our communicator ray can't get through that mirror, but it looks as though our best bet is to act peaceable and see if we can't talk to them in some way. Right?"

"Right." They stepped out into the airlock, from which they saw that the great sphere had halted only a few yards from them, and that an indistinct figure stood in an open door, waving to them an unmistakable invitation to enter the strange vessel.

"Shall we, Steve?"

"Might as well. They've got us foul, and can take us if they want us. Anyway, we'll need at least a week to fix us up any kind of driving power, so we can't run—and we probably couldn't get away from those folks if we had all our power. They haven't blown us up, and they could have done it easily enough. Besides, they act friendly, so we'd better meet them half way. Dive!"

Floating toward the open doorway, they were met by another rod of force, brought gently into the airlock, and supported upright beside the being who had invited them to visit him. Apparently an empty space-suit stood there; a peculiarly-fitted suit of some partially transparent, flexible, glass-like material; towering fully a foot over the head of the tall Terrestrial. Closer inspection, however, revealed that there was something inside that suit—a shadowy, weirdly-transparent being, staring at them with large, black eyes. The door clanged shut behind them; they heard the faint hiss of inrushing air, and the inner door opened; but their enveloping suits remained stretched almost as tightly as ever. They felt the floor lurch beneath their feet, and a little weight was granted them as the space-ship got under way. Stevens waved his arms vigorously at the stranger, pointing backward toward where he supposed their own craft to be. The latter waved an arm reassuringly, pressed a contact, and a section of the wall suddenly became transparent. Through it Stevens saw with satisfaction that the *Forlorn Hope* was not being abandoned; in the grip of powerful tractor beams, every fragment of the wreckage was following close behind them in their flight through space.

STEVENS and Nadia followed their guide along a corridor, through several doors, and into a large room, which at first glance seemed empty, but in which several of the peculiarly transparent people of the craft were lying about upon cushions. They were undoubtedly human—but what humans! Tall and reedy they were, with enormous barrel chests, topped by heads which, though really large, appeared insignificant because of the prodigious chests and because of the huge, sail-like, flapping ears. Their skins were a strikingly, livid, pale blue, absolutely devoid of hair; and their lidless eyes, without a sign of iris, were chillingly horrible in their stark contrast of enormous, glaring black pupil and ghastly, transparent blue eyeball.

As the two Terrestrials entered the room, the beings struggled to their feet and hurried laboriously away. Soon one of them returned, dressed in an insulating suit, and carrying three sets of head harnesses, connected by multiplex cables to a large box which he placed upon the floor. He handed the headsets to the first officer, who in turn placed two of them at the feet of the Terrestrials, indicating to them that they were to follow his example in placing them upon their heads, outside the helmets. They did so, and even through the almost perfect insulation, and in spite of the powerful heaters of their suits, they felt a touch of frightful cold. The stranger turned a dial, and the two wanderers from Earth were instantly in full mental communication with Barkovis, the commander of a space-ship of Titan, the sixth satellite of Saturn!

"Well, I'll be ... say, what is this, anyway?" Steve exclaimed involuntarily, and Nadia smiled as Barkovis answered with a thought, clearer than any spoken words.

"It is a thought-exchanger. I do not know its fundamental mechanism, since we did not invent it and since I have had little time to study it. The apparatus, practically as you see it here, was discovered but a short time ago, in a small, rocket-propelled space-ship which we found some distance outside of the orbit of Jupiter. Its source of power had been destroyed by the cold of outer space, but re-powering it was, of course, a small matter. The crew of the vessel were all dead. They were, however, of human stock, and of a type adapted for life upon a satellite. I deduce, from your compact structure, your enormous atmospheric pressure, and your, to us, unbelievably high body temperature, that you must be planet-dwellers. I suppose that you are natives of Jupiter?"

"Not quite." Stevens had in a measure recovered from his stunned surprise. "We are from Tellus, the third planet," and he revealed rapidly the events leading up to their present situation, concluding: "The people in the other sphere were, we believe, natives of Jupiter or of one of the satellites. We know nothing of them, since we could not look through their screens. You rescued us from them; do you not know them?"

"No. Our visirays also were stopped by their screens of force—screens entirely foreign to our science. This is the first time that any vessel from our Saturnian system has ever succeeded in reaching the neighborhood of Jupiter. We came in peace, but they attacked us at sight and we were obliged to destroy them. Now we must hurry back to Titan, for two reasons. First, because we are already at the extreme limit of our power range and Jupiter is getting further and further away from Saturn. Second because our mirrors, which we had thought perfect reflectors of all frequencies possible of generation, are not perfect. Enough of those forces came through the mirrors to volatilize half our crew, and in a few minutes more none of us would have been left alive. Why, in some places our very atmosphere became almost hot enough to melt water! If another of those vessels should attack us, in all probability we should all be lost. Therefore we are leaving as rapidly as is possible."

"You are taking the pieces of our ship along—we do not want to encumber you."

"It is no encumbrance, since we have ample supplies of power. In fact, we are now employing the highest acceleration we Titanians can endure for any length of time."

Stevens pondered long, forgetting that his thoughts were plain as print to the Titanian commander. Thank Heaven these strangers had sense enough to be friendly—all intelligent races should be friends, for mutual advancement. But it was a mighty long stretch to Saturn and this acceleration wasn't so much. How long would it take to get there? Could they get back? Wouldn't they save time by casting themselves adrift, making the repairs most urgently needed, and going back to Ganymede under their own power? But would they have enough power left in the wreck to get even that far? And how about the big tube? He was interrupted by an insistent thought from Barkovis.

"You will save time, Stevens, by coming with us to Titan. There we shall aid you in repairing your vessel and in completing your transmitting tube, in which we shall be deeply interested. Our power plants shall supply you with energy for your return journey until you are close enough to Jupiter to recover your own beam. You are tired. I would suggest that you rest—that you sleep long and peacefully."

"You seem to be handling the *Forlorn Hope* without any trouble—the pieces aren't grinding at all. We'd better live there, hadn't we?"

"Yes that would be best, for all of us. You could not live a minute here without your suits; and, efficiently insulated as those suits are, yet your incandescent body temperature makes our rooms unbearably hot—so hot that any of us must wear a space-suit while in the same room with you, to avoid being burned to death."

"The incandescently hot" Terrestrials were wafted into the open airlock of their lifeboat upon a wand of force, and soon had prepared a long overdue supper, over which Stevens cast his infectious, boyish grin at Nadia.

"Sweetheart, you are undoubtedly a 'warm number,' and you have often remarked that I 'burn you up.' Nevertheless I think that we were both considerably surprised to discover that we are both hot enough actually to consume persons unfortunate enough to be confined in the same room with us!"

"You're funny, Steve—like a crutch," she rebuked him, but smiled back, an elusive dimple playing in one lovely brown cheek. "Looking right through anybody is too ghastly for words, but I think they're perfectly all x, anyway, in spite of their being so hideous and so cold-blooded!"

CHAPTER VI
A Frigid Civilization

"HI, Percival Van Schravendyck Stevens!" Nadia strode purposely into Stevens' room and seized him by the shoulder. "Are you going to sleep all the way to Saturn? You answered me when I pounded on the partition with a hammer, but I don't believe that you woke up at all. Get up, you—breakfast will be all spoiled directly!"

"Huh?" Stevens opened one sluggish eye; then, as the full force of the insult penetrated his consciousness, he came wide awake. "Lay off those names, ace, or you'll find yourself walking back home!" he threatened.

"All x by me!" she retorted. "I might as well go home if you're going to sleep *all* the time!" and she widened her expressive eyes at him impishly as she danced blithely back into the control room. As she went out she slammed his door with a resounding clang, and Stevens pried himself out of his bunk one joint at a time, dressed, and made himself presentable.

"Gosh!" he yawned mightily as he joined the girl at breakfast. "I don't know when I've had such a gorgeous sleep. How do you get by on so little?"

"I don't. I sleep a lot, but I do it every night, instead of working for four days and nights on end and then trying to make up all those four nights' sleep at once. I'm going to break you of that, too, Steve, if it's the last thing I ever do."

"There might be certain advantages in it, at that," he conceded, "but sometimes you've got to do work when it's got to be done, instead of just between sleeps. However, I'll try to do better. Certainly it is a wonderful relief to get out of that mess, isn't it?"

"I'll say it is! But I wish that those folks were more like people. They're nice, I think, really, but they're so ... so ... well, so ghastly that it simply gives me the blue shivers just to look at one of them!"

"They're pretty gruesome, no fooling," he agreed, "but you get used to things like that. I just about threw a fit the first time I ever saw a Martian, and the Venerians are even worse in some ways—they're so clammy and dead-looking—but now I've got real friends on both planets. One thing, though, gives me the pip. I read a story a while ago—the latest best-seller thing of Thornton's named 'Interstellar Slush' or some such tr...."

"Cleophora—An Interstellar Romance," she corrected him. "I thought it was wonderful!"

"I didn't. It's fundamentally unsound. Look at our nearest neighbors, who probably came from the same original stock we did. A Tellurian can admire, respect, or like a Venerian, yes. But for *loving* one of them—wow! Beauty is purely relative, you know. For instance, I think that you are the most perfectly beautiful thing I ever saw; but no Venerian would think so. Far from it. Any Martian that hadn't seen many of us would have to go rest his eyes after taking one good look at you. Considering what love means, it doesn't stand to reason that any Tellurian woman could possibly fall in love with any man not of her own breed. Any writer is wrong who indulges in interplanetary love affairs and mad passions. They simply don't exist. They *can't* exist—they're against all human instincts."

"Interplanetary—in this solar system—yes. But the Dacrovos were just like us, only nicer."

"That's what gives me the pip. If our own cousins of the same solar system are so repulsive to us, how would we be affected by entirely alien forms of intelligence?"

"May be you're right, of course—but you may be wrong, too," she insisted. "The Universe is big enough, so that people like the Dacrovos may possibly exist in it somewhere. May be the Big Three will discover a means of interstellar travel—then I'll get to see them myself, perhaps."

"Yes, and *if* we do, and *if* you ever see any such people, I'll bet that the sight of them will make your hair curl right up into a ball, too! But about Barkovis—remember how diplomatic the thoughts were that he sent us? He described our structure as being 'compact,' but I got the undertone of his real thoughts, as well. Didn't you?"

"Yes, now that you mention it, I did. He really thought that we were white-hot, under-sized, overpowered, warty, hairy, hideously opaque and generally repulsive little monstrosities—thoroughly unpleasant and distasteful. But he was friendly, just the same. Heavens, Steve! Do you suppose that he read our real thoughts, too?"

"Sure he did; but he is intelligent enough to make allowances, the same as we are doing. He isn't any more insulted than we are. He knows that such feelings are ingrained and cannot be changed."

BREAKFAST over, they experienced a new sensation. For the first time in months they had nothing to do! Used as they were to being surrounded by pressing tasks, they enjoyed their holiday immensely for a few hours. Sitting idly at the communicator plate, they scanned the sparkling heavens with keen interest. Beneath them Jupiter was a brilliant crescent not far from the sun in appearance, which latter had already grown perceptibly smaller and less bright. Above them, and to their right, Saturn shone refulgently, his spectacular rings plainly visible. All about them were the glories of the firmament, which never fail to awe the most seasoned observer. But idleness soon became irksome to those two active spirits, and Stevens prowled restlessly about their narrow quarters.

"I'm going to go to work before I go dippy," he soon declared. "They've got lots of power, and we can rig up a transmitter unit to send it over here to our receptor. Then I can start welding the old *Hope* together without waiting until we get to Titan to start it. Think I'll signal Barkovis to come over, and see what he thinks about it."

The Titanian commander approved the idea, and the transmitting field was quickly installed. Nadia insisted that she, too, needed to work, and that she was altogether too good a mechanic to waste; therefore the two again labored mightily together, day after day. But the girl limited rigidly their hours of work to those of the working day; and evening after evening Barkovis visited with them for hours. Dressed in his heavy space-suit and supported by a tractor beam well out of range of what seemed to him terrific heat radiated by the bodies of the Terrestrials, he floated along unconcernedly; while over the multiplex cable of the thought-exchanger he conversed with the man and woman seated just inside the open outer door of their air-lock. The Titanian's appetite for information was insatiable—particularly did he relish everything pertaining to the earth and to the other inner planets, forever barred to him and to his kind. In return Stevens and Nadia came gradually to know the story of the humanity of Titan.

"I am glad beyond measure to have known you," Barkovis mused, one night. "Your existence proves that there is truth in mythology, as some of us have always believed. Your visit to Titan will create a furor in scientific circles, for you are impossibility incarnate—personifications of the preposterous. In you, wildest fancy had become commonplace. According to many of our scientists, it is utterly impossible for you to exist. Yet you say, and it must be, that there are millions upon millions of similar beings. Think of it! Venerians, Tellurians, Martians, the satellite dwellers of the lost space-ship, and us—so similar mentally, yet physically how different!"

"But where does the mythology come in?" thought Nadia.

"We have unthinkably ancient legends which say that once Titan was extremely hot, and that our remote ancestors were beings of fire, in whose veins ran molten water instead of blood. Since our recorded history goes back some tens of thousands of Saturnian years, and since in that long period there has been no measurable change in us, few of us have believed in the legends at all. They have been thought the surviving figments of a barbarous, prehistoric worship of the sun. However, such a condition is not in conflict with the known facts of cosmogony, and since there actually exists such a humanity as yours—a humanity whose bodily tissues actually *are* composed largely of molten water—those ancient legends must indeed have been based upon truth.

"What an evolution! Century after century of slowly decreasing temperature—one continuous struggle to adapt the physique to a constantly changing environment. First they must have tried to maintain their high temperature by covering and heating their cities.—Then, as vegetation died, they must have bred into their plants the ability to use as sap purely chemical liquids, such as our present natural fluids—which also may have been partly synthetic then—instead of the molten water to which they had been accustomed. They must have modified similarly the outer atmosphere; must have made it more reactive, to compensate for the lowered temperature at which metabolism must take place. As Titan grew colder and colder they probably dug their cities deeper and ever deeper; until humanity came finally to realize that it must itself change completely or perish utterly.

"Then we may picture them as aiding evolution in changing their body chemistry. For thousands, and thousands of years there must have gone on the gradual adaptation of blood stream and tissue to more and more volatile liquids, and to lower and still lower temperatures. This must have continued until Titan arrived at the condition which has now obtained for ages—a condition of thermal equilibrium with space upon one hand and upon the other the sun, which changes appreciably only in millions upon millions of years. In equilibrium at last—with our bodily and atmospheric temperatures finally constant at their present values, which seem as low to you as yours appear high to us. Truly, an evolution astounding to contemplate!"

"But how about power?" asked Stevens. "You seem to have all you want, and yet it doesn't stand to reason that there could be very much generated upon a satellite so old and so cold."

"You are right. For ages there has been but little power produced upon Titan. Many cycles ago, however, our scientists had developed rocket-driven space-ships, with which they explored our neighboring satellites, and even Saturn itself. It is from power plants upon Saturn that we draw energy. Their construction was difficult in the extreme, since the pioneers had to work in braces because of the enormous force of gravity. Then, too, they had to be protected from the overwhelming pressure and poisonous qualities of the air, and insulated from a temperature far above the melting point of water. In such awful heat, of course, our customary building material, water, could not be employed...."

"But all our instruments have indicated that Saturn is *cold*!" Stevens interrupted.

"Its surface temperature, as read from afar, would be low," conceded Barkovis, "but the actual surface of the planet is extremely hot, and is highly volcanic. Practically none of its heat is radiated because of the great density and depth of its atmosphere, which extends for many hundreds of your kilometers. It required many thousands of lives and many years of time to build and install those automatic power plants, but once they were in operation, we were assured of power for many tens of thousands of years to come."

"Our system of power transmission is more or less like yours, but we haven't anything like your range. Suppose you'd be willing to teach me the computation of your fields?"

"Yes, we shall be glad to give you the formulae. Being an older race, it is perhaps natural that we should have developed certain refinements as yet unknown to you. But I am, I perceived, detaining you from your time of rest—goodbye," and Barkovis was wafted back toward his mirrored globe.

"What do you make of this chemical solution blood of theirs, Steve?" asked Nadia, watching the placidly floating form of the Titanian captain.

"Not much. I may have mentioned before that there are one or two, or perhaps even three men who are better chemists than I am. I gathered that it is something like a polyhydric alcohol and something like a substituted hydrocarbon, and yet different from either in that it contains flourin in loose combination. I think it is something that our Tellurian chemists haven't got yet; but they've got so many organic compounds now that they may have synthesized it, at that. You see, Titan's atmosphere isn't nearly as dense as ours, but what there is of it is pure dynamite. Ours is a little oxygen, mixed with a lot of inert ingredients. Theirs is oxygen, heavily laced with flourin. It's *reactive*, no fooling! However, something pretty violent must be necessary to carry on body reactions at such a temperature as theirs."

"Probably; but I know even less about that kind of thing than you do. Funny, isn't it, the way he thinks 'water' when he means ice, and always thinks of our real water as being molten?"

"Reasonable enough when you think about it. Temperature differences are logarithmic, you know, not arithmetic—the effective difference between his body temperature and ours is perhaps even greater than that between ours and that of melted iron. We never think of iron as being a liquid, you know."

"That's right, too. Well, good night, Steve dear."

"'Bye, little queen of space—see you at breakfast," and the *Forlorn Hope* became dark and silent.

DAY after day the brilliant sphere flew toward distant Saturn, with the wreckage of the *Forlorn Hope* in tow. Piece by piece that wreckage was brought together and held in place by the Titanian tractors; and slowly bu steadily, under Stevens' terrific welding projector, the stubborn steel flow together, once more to become a seamless, spaceworthy structure. A Nadia, the electrician, followed close behind the welder. Wielding tor pliers and spanner with practised hand, she repaired or cut out of circ the damaged accumulator cells and reunited the ends of each sever power lead. Understanding Nadia's work thoroughly, the Titanians we not particularly interested in it; but whenever Stevens made his way alor an outside seam, he had a large and thrillingly horrified gallery. Everyor who could possibly secure permission to leave the sphere did so, each upc his own pencil of force, and went over to watch the welder. They did n come close to him—to venture within fifty feet of that slow moving spot scintillating brilliance, even in a space-suit, meant death—but, poised arou him in space, they watched with shuddering, incredulous amazement, t monstrous human being in whose veins ran molten water instead of bloo whose body was already so fiercely hot that it could exist unharmed whi working practically without protection, upon *liquefied* metal!

Finally the welding was done. The insulating space was evacuated an held its vacuum—outer and inner shells were bottle-tight. The two mechanic heaved deep sighs of relief as they discarded their cumbersome armor an began to repair what few of their machine tools had been damaged by th slashing plane of force which had so neatly sliced the *Forlorn Hope* int sections.

"Say, big fellow, you're the guy that slings the ink, ain't you?" Nadia extinguished her torch and swaggered up to Stevens, hands on hips, he walk an exaggerated roll. "Write me out a long walk. This job's all playe out, so I think I'll get me a good job on Titan. I said give me my time, y big stiff!"

"You didn't say nothing!" growled Stevens in his deepest bass, playi up to her lead as he always did. "Bounce back, cub, you've struck a rubb fence! You signed on for duration and you'll stick—see?"

Arm in arm they went over to the nearest communicator plate. Flipping the switch, Stevens turned the dial and Titan shone upon the screen; sc close, that it no longer resembled a moon, but was a world toward which they were falling with an immense velocity.

"Not close enough to make out much detail yet—let's take another look at Saturn," and Stevens projected the visiray beam out toward the mighty planet. It was now an enormous full moon, almost five degrees in apparent diameter,[1] its visible surface an expanse of what they knew to be billowing cloud, shining brilliantly white in the pale sunlight, broken only by a dark equatorial band.

"Those rings were *such* a gorgeous spectacle a little while ago!" Nadia mourned. "It's a shame that Titan has to be right in their plane, isn't it? Think of living this close to one of the most wonderful sights in the Solar System, and never being able to see it. Think they know what they're missing, Steve?"

"We'll have to ask Barkovis," Stevens replied. He swung the communicator beam back toward Titan, and Nadia shuddered.

"Oh, it's hideous!" she exclaimed. "I thought that it would improve as we got closer, but the plainer we can see it, the worse it gets. Just to think of human beings, even such cold-blooded ones as those over there, living upon such a horrible moon and *liking* it, gives me the blue shivers!"

"It's pretty bleak, no fooling," he admitted, and peered through the eyepiece of the visiray telescope, studying minutely the forbidding surface of the satellite they were so rapidly approaching.

Larger and larger it loomed, a cratered, jagged globe of desolation indescribable; of sheer, bitter cold incarnate and palpable; of stark, sharp contrasts. Gigantic craters, in whose yawning depths no spark of warmth had been generated for countless cycles of time, were surrounded by vast plains eroded to the dead level of a windless sea. Every lofty object cast a sharply outlined shade of impenetrable blackness, beside which the weak light of the sun became a dazzling glare. The ground was either a brilliant white or an intense black, unrelieved by half-tones.

"I can't hand it much, either, Nadia, but it's all in the way you've been brought up, you know. This is home to them, and just to look at Tellus would give them the pip. Ha! Here's something you'll like, even if it does look so cold that it makes me feel like hugging a couple of heater coils. It's Barkovis' city the one we're heading for, I think. It's close enough now so that we can get it on the plate," and he set the communicator beam upon the metropolis of Titan.

"Why, I don't see a thing, Steve—where and what is it?" They were dropping vertically downward toward the center of a vast plain of white, featureless and desolate; and Nadia stared in disappointment.

"You'll see directly—it's too good to spoil by telling you what to look for or wh...."

"Oh, there it is!" she cried. "It *is* beautiful, Steve, but how frightfully, utterly cold!"

AFLASH of prismatic color had caught the girl's eye, and, one transparent structure thus revealed to her sight, there had burst into view a city of crystal. Low buildings of hexagonal shape, arranged in irregularly variant hexagonal patterns, extended mile upon mile. From the roofs of the structures lacy spires soared heavenward; inter-connected by long, slim cantilever bridges whose prodigious spans seemed out of all proportion to the gossamer delicacy of their construction. Buildings, spires, and bridges formed fantastic geometrical designs, at which Nadia exclaimed in delight.

"I've just thought of what that reminds me of—it's snowflakes!"

"Sure—I knew it was something familiar. Snowflakes—no two are ever exactly alike, and yet every one is symmetrical and hexagonal. We're going to land on the public square—see the crowds? Let's put on our suits and go out."

The *Forlorn Hope* lay in a hexagonal park, and near it the Titanian globe had also come to rest. All about the little plot towered the glittering buildings of crystal, and in its center played a fountain; a series of clear and sparkling cascades of liquid jewels. Under foot there spread a thick, soft carpet of whitely brilliant vegetation. Throngs of the grotesque citizens of Titania were massed to greet the space-ships; throngs clustering close about the globular vessel, but maintaining a respectful distance from the fiercely radiant Terrestrial wedge. All were shouting greetings and congratulations—shouts which Stevens found as intelligible as his own native tongue.

"Why, I can understand every word they say, Steve!" Nadia exclaimed, in surprise. "How come, do you suppose?"

"I can, too. Don't know—must be from using that thought telephone of theirs so much, I guess. Here comes Barkovis—I'll ask him."

The Titanian commander had been in earnest conversation with a group of fellow-creatures and was now walking toward the Terrestrials, carrying the multiple headsets. Placing them upon the white sward, he backed away, motioning the two visitors to pick them up.

"It may not be necessary, Barkovis," Stevens said, slowly and clearly. "We do not know why, but we can understand what your people are saying, and it may be that you can now understand us."

"Oh, yes, I can understand your English perfectly. A surprising development, but perhaps, after all, one that should have been expected, from the very nature of the device we have been using. I wanted to tell you that I have just received grave news, which makes it impossible for us to help you immediately, as I promised. While we were gone, one of our two power-plants upon Saturn failed. In consequence, Titan's power has been cut to a minimum, since maintaining our beam at that great distance required a large fraction of the output of the other plant. Because of this lack, the Sedlor walls were weakened to such a point that in spite of the Guardian's assurances, I think trouble is inevitable. At all events, it is of the utmost importance that we begin repairing the damaged unit, for that is to be a task indeed."

"Yes, it will take time," agreed Stevens, remembering what the Titanian captain had told him concerning the construction of those plants—generators which had been in continuous and automatic operation for thousands of Saturnian years.

"It will take more than time—it will take lives," replied Barkovis, gravely. "Scores, perhaps hundreds, of us will never again breathe the clear, pure air of Titan. In spite of all precaution and all possible bracing and insulation, man after man after man will be crushed by his own weight, volatilized by the awful heat, poisoned by the foul atmosphere, or will burst into unthinkable flames at the touch of some flying spark from the inconceivably hot metals with which we shall have to work. A horrible fate, but we shall not lack for volunteers."

"Sure not; and of course you yourself would go. And I never thought of the effect a spark would have on you—your tissues would probably be wildly inflammable. But say, I just had a thought. Just how hot is the air at those plants and just what is the actual pressure?"

"According to the records, the temperature is some forty of your centigrade degrees above the melting-point of water, and the pressure is not far short of two of your meters of mercury. I find it almost impossible to think of mercury as a liquid, however."

"You find it impossible, since you use it as a metal, for wires in coils and so on. But plus forty, while pretty warm, isn't impossible, by any means; and we could stand double our air pressure for quite a while. Both my partner and I are pretty fair mechanics and we've got quite a line of machine tools, such as you could not possibly have here. We'll give it a whirl, since we owe you something already. Lead us to it, ace—but wait a minute! We can't see through the fog, so couldn't find the plants, and probably your wiring diagrams would explode if I touched them."

"I never thought of your helping us," mused Barkovis. "The idea of any living being existing in that inferno has always been unthinkable, but the difficulties you mention are slight. We have already built in our vessel communicators similar to yours, and radio sets. With these we can guide you and explain the plants to you as you work, and our tractor beams will be of assistance to you in moving heavy objects, even at such distances from the surface as we Titanians shall have to maintain. If you will set out a flask of your atmosphere, we will analyze it, for the thought has come to me that perhaps, being planet-dwellers yourselves, the air of Saturn might not be as poisonous to you as it is to us."

"That's a thought, too," and, the news broadcast, it was not long until the two ships leaped into the air, to the accompaniment of the cheers and plaudits of a watching multitude.

IN a wide curve they sped toward Saturn. Passing so close to the enormous rings that the individual meteoric fragments could almost be seen with the unaided eye, they flashed on and on, slowing down long before they approached the upper surface of the envelope of cloud. The spherical space-ship stopped and Stevens, staring into his useless screen, drove the *Forlorn Hope* downward mile after mile, solely under Barkovis' direction, changing course and power from time to time as the Titanian's voice came from the speaker at his elbow. Slower and slower became the descent, until finally, almost upon the broad, flat roof of the power-plant, Stevens saw it in his plate. Breathing deeply in relief, he dropped quickly down upon a flat pavement, neutralized his controls, and turned to Nadia.

"Well, old golf-shootist, we're here at last—now we'll go out and see what's gone screwy with the works. Remember that gravity is about double normal here, and conduct yourself accordingly."

"But it's supposed to be only about nine-tenths," she objected.

"That's at the outer surface of the atmosphere," he replied. "And it's *some* atmosphere—not like the thin layer we've got on Tellus."

They went into the airlock, and Stevens admitted air until their suits began to collapse. Then, face-plate valves cracked, he sniffed cautiously, finally opening his helmet wide. Nadia followed suit and the man laughed as she wrinkled her nose in disgust as two faint, but unmistakable odors smote her olfactory nerves.

"I never cared particularly for hydrogen sulphide and sulphur dioxide, either," he assured her, "but they aren't strong enough to hurt us in the short time we'll be here. Those Titanian chemists know their stuff, though."

He opened the outer valves slowly, then opened the door and they stepped down upon the smooth, solid floor, which Stevens examined carefully.

"I thought so, from his story. Solid platinum! This whole plant is built of platinum, iridium, and noble alloys—the only substances known that will literally last forever. Believe me, ace of my bosom, I don't wonder that it cost them lives to build it—with their conditions, I don't see how they ever got it built at all."

Before them rose an immense, flat-topped cone of metal, upon the top of which was situated the power plant. Twelve massive pillars supported a flat roof, but permitted the air to circulate freely throughout the one great room which housed the machinery. They climbed a flight of stairs, passed between two pillars, and stared about them. There was no noise, no motion—there was nothing that *could* move. Twelve enormous masses of metallic checkerwork, covered with wide cooling fins, almost filled the vast hall. From the center of each mass great leads extended out into a clear space in the middle of the room, there uniting in mid-air to form one enormous bus-bar. This bar, thicker than a man's body, had originally curved upward to the base of an immense parabolic structure of latticed bars. Now, however, it was broken in midspan and the two ends bent toward the floor. Above their heads, a jagged hole gaped in the heavy metal of the roof, and a similar hole had been torn in the floor. The bar had been broken and these holes had been made by some heavy body, probably a meteorite, falling with terrific velocity.

"This is it, all right," Stevens spoke to distant Barkovis. "Sure there's nothing on this beam? If it should be hot and I should short circuit or bridge it with my body, it would be just too bad."

"We have made sure that nothing is connected to it," the Titanian assured him. "Do you think you can do anything?"

"Absolutely. We've got jacks that'll bend heavier stuff than that, and after we get it straightened the welding will be easy, but I'll have to have some metal. Shall I cut a piece off the pavement outside?"

"That will not be necessary. You will find ample stores of spare metal piled at the base of each pillar."

"All x. Now we'll get the jack, Nadia," and they went back to their vessel, finding that upon Saturn, their combined strength was barely sufficient to drag the heavy tool along the floor.

"Stand aside, please. We will place it for you," a calm voice sounded in their ears, and a pale blue tractor beam picked the massive jack lightly from the floor, and as lightly lifted it to its place beneath the broken bus-bar and held it there while Stevens piled blocks and plates of platinum beneath its base.

"Well, here's where I peel down as far as the law allows. This is going to be real work, girl—no fooling. It'd help a lot if this outfit were sending out a few thousand kilo-franks instead of standing idle."

"How would that help?"

"It's a heat-engine, you know—works by absorbing heat. The cold air sinks—I imagine it pretty nearly blows a gale down the side of this cone when it's working—and hot air rushes in to take its place. I could use a little cool breeze right now," and Stevens, stripped to the waist, bent to the lever of the powerful hydraulic jack.

Beads of sweat gathered upon his broad back, uniting to form tiny rivulets, and the girl became highly concerned about him.

"Let me help you, Steve—I'm pretty husky, too, you know."

"Sure you are, ace, but this is a job for a truck-horse, not a tenderly-nurtured maiden of the upper classes. You can help, though, by breaking out that welding outfit and getting it ready while I'm doing this bending to prepare for the welding."

Under the urge of that mighty jack the ends of the broken bus-bar rose into place, while far off in space the Titanians clustered about their visiray screens, watching, in almost unbelieving amazement, the supernatural being who labored in that reeking inferno of heat and poisonous vapor—who labored almost naked and entirely unprotected, refreshing himself from time to time with drafts of molten water!

"All x, Barkovis—that's high, I guess." Stevens flipped perspiration from his hot forehead with a wet finger and straightened his weary back. "Now you can put this jack away where we had it. Then you might trundle me over enough of that spare metal to fill up this hole, and I'll put on my suit and goggles and practice welding on this floor and the roof, to get the feel of the metal before I tackle the bar."

The hole in the floor was filled with scrap and soon sparks were flying wildly as the searing beam of Stevens' welding projector bit viciously into the stubborn alloy of noble metals; fashioning a smooth, solid floor where the yawning aperture had been. Then, lifted with his tools and plates to the roof, the man repaired that hole also.

"Now I know enough about it to do a good job on the bar," he decided, and brick after brick of alloy was fused into the crack, until only a smoothly rounded bulge betrayed that a break had ever existed in that mighty rod of metal.

"Give 'em the signal to draw power, and see if that's all that was the matter," Stevens instructed, as he relaxed in the grateful coolness of their control room. "Whew, that was a warm job, Nadia—and this air of ours does smell good!"

"IT was a horrible job, and I'm glad it's done," she declared. "But say, Steve, that thing looks as little like a power-plant as anything I can imagine. How does it work? You said that it worked on heat, but I don't quite see how. But don't draw diagrams and *please* don't integrate!"

"No ordinary plant such as we use could run for centuries without attention," he replied. "This is a highly advanced heat-engine—something like a thermo-couple, you know. This whole thing is simply the hot end, connected to the cold end on Titan by a beam instead of wires. When it's working, this metal must cool off something fierce. That's what the checkerwork and fins are for—so that it can absorb the maximum amount of heat from the current of hot, moist air I spoke about. It's a sweet system—we'll have to rig up one between Tellus and the moon. Or even between the Equator and the Arctic Circle there'd be enough thermal differential to give us a million kilofranks. We haven't got the all x signal yet, but it's working—look at it sweat as it cools down!"

"I'll say it's sweating—the water is simply streaming off it!" In their plate they saw that moisture was already beginning to condense upon the heat-absorber: moisture running down the fins in streams and creeping over the dull metal floor in sluggish sheets; moisture which, turning into ice in the colder interior of the checkerwork, again became fluid at the inrush of hot, wet Saturnian air.

"There's the signal—all x, Barkovis? By the way it's condensing water, it seems to be functioning again."

"Perfect!" came the Titanian's enthusiastic reply, "You two planet-dwellers have done more in three short hours than the entire force of Titan could have accomplished in months. You have earned, and shall receive, the highest...."

"As you were, ace!" Stevens interrupted, embarrassed. "This job was just like shooting fish down a well, for us. Since you saved our lives, we owe you a lot yet. We're coming out—straight up!"

The *Forlorn Hope* shot upward, through mile after mile of steaming fog, until at last she broke through into the light, clear outer atmosphere. Stevens located the Titanian space-ship, and the two vessels once more hurtling together through the ether toward Titan, he turned to his companion.

"Take the controls, will you, Nadia? Think I'll finish up the tube. I brought along a piece of platinum from the power plant, and something that I think is tantalum from Barkovis' description of it. With those and the fractions we melted out, I think I can make everything we'll need."

Now that he had comparatively pure metal with which to work, drawing the leads and filaments was relatively a simple task. Working over the hot-bench with torch and welding projector, he made short work of running the leads through the almost plastic glass of the great tube and of sealing them in place. The plates and grids presented more serious problems; but they were solved and, long before Titan was reached, the tube was out in space, supported by a Titanian tractor beam between the two vessels. Stevens came into the shop, holding a modified McLeod gauge which he had just taken from the interior of the tube. When it had come to equilibrium, he read it carefully and yelled.

"Eureka, little fellow! She's down to where I can't read it, even on this big gauge—so hard that it won't need flashing—harder than any vacuum I ever got on Tellus, even with a Rodebush-Michalek super-pump!"

"But how about occluded and absorbed gas in the filaments and so on when they heat up?" demanded Nadia, practically.

"All gone, ace. I out-gassed 'em plenty out there—seven times, almost to fusion. There isn't enough gas left in the whole thing to make a deep breath for a microbe."

He took up his welding projector and a beam carried him back to the tube. There, in the practically absolute vacuum of space, the last openings in the glass were sealed, and man and great transmitting tube were wafted lightly back into the Terrestrial cruiser.

Hour after hour mirrored Titanian sphere and crude-fashioned terrestrial wedge bored serenely on through space, and it was not until Titan loomed large beneath them that the calm was broken by an insistent call from Titan to the sphere.

"Barkodar, attention! Barkodar, attention!" screamed from the speakers, and they heard Barkovis acknowledge the call.

"The Sedlor have broken through and are marching upon Titania. The order has gone out for immediate mobilization of every unit."

"There's that word 'Sedlor' again—what are they, anyway, Steve?" demanded Nadia.

"I don't know. I was going to ask him when he sprung it on us first, but he was pretty busy then and I haven't thought of it since. Something pretty serious, though—they've jumped their acceleration almost to Tellurian gravity, and none of them can live through much of that."

"Tellurians?" came the voice of Barkovis from the speaker. "We have just...."

"All x—we were on your wave and heard it," interrupted Stevens. "We're with you. What are those Sedlor, anyway? Maybe we can help you dope out something."

"Perhaps—but whatever you do, do not use your heat-projector. That would start a conflagration raging over the whole country, and we shall have enough to do without fighting fire. But it may be that you have other weapons, of which we are ignorant, and I can use a little time in explanation before we arrive. The Sedlor are a form of life, something like your..." he paused, searching through his scanty store of Earthly knowledge, then went on, doubtfully, "perhaps some thing like your insects. They developed a sort of intelligence, and because of their fecundity, adapted themselves to their environment as readily as did man; and for ages they threatened man's supremacy upon Titan. They devoured vegetation, crops, animals, and mankind. After a world-wide campaign, however, they were finally exterminated, save in the neighborhood of one great volcanic crater, which they so honeycombed that it is almost impregnable. All around that district we have erected barriers of force, maintained by a corps of men known as 'Guardians of the Sedlor.' These barriers extend so far into the ground and so high into the air that the Sedlor can neither burrow beneath them nor fly over them. They were being advanced as rapidly as possible, and in a few more years the insects would have been destroyed completely—but now they are again at large. They have probably developed an armor or a natural resistance greater than the Guardians thought possible, so that when the walls were weakened, they came through in their millions, underground and undetected. They are now attacking our nearest city—the one you know, and which you have called Titania."

"What do you use—those high-explosive bombs?"

"The bombs were developed principally for use against them, but proved worse than useless, for we found that when a Sedlor was blown to pieces, each piece forthwith developed into a new, complete creature. Our most efficient weapons are our heat rays—not yours remember—and poison gas. I must prepare our arms."

"Would our heat-ray actually set them afire, Steve?" Nadia asked, as the plate went blank.

"I'll say it would. I'll show you what heat means to them—showing you will be plainer than any amount of explanation," and he shot the visiray beam down toward the city of Titania. Into a low-lying building it went, and Nadia saw a Titanian foundry in full operation. Men clad in asbestos armor were charging, tending, and tapping great electric furnaces and crucibles; shrinking back and turning their armored heads away as the hissing, smoking melt crackled into the molds from their long-handled ladles. Nadia studied the foundry for a moment, interested, but unimpressed.

"Of course it's hot there—foundries always *are* hot," she argued.

"Yes, but you haven't got the idea yet." Stevens turned again to the controls, following the sphere toward what was evidently a line of battle. "That stuff that they are melting and casting and that is so hot, is not metal, but *ice!* Remember that the vital fluid of all life here, animal and vegetable, corresponding to our water, is probably more inflammable than gasoline. If they can't work on ice-water without wearing suits of five-ply asbestos, what would a real heat-ray do to them? It'd be about like our taking a dive into the sun!"

"*Ice*!" she exclaimed. "Oh of course—but you couldn't really believe a thing like that without seeing it, could you? Oh, Steve—how utterly horrible!"

THE "Barkodar" had dropped down into a line of sister ships, and had gone into action in midair against a veritable swarm of foes. Winged centipedes they were—centipedes fully six feet long, hurling themselves along the ground and through the air in furious hordes. From the flying globes emanated pale beams of force, at the touch of which the Sedlor disappeared in puffs of vapor. Upon the ground huge tractors and trucks, manned by masked soldiery, mounted mighty reflectors projecting the same lethal beam. From globes and tanks there sounded a drumming roar and small capsules broke in thousands among the foe; emitting a red cloud of gas in which the centipedes shriveled and died. But for each one that was destroyed two came up from holes in the ground and the battle-line fell back toward Titania, back toward a long line of derrick-like structures which were sinking force-rods into the ground in furious haste.

Stevens flashed on his ultra-violet projector and swung it into the thickest ranks of the enemy. In the beam many of the monsters died, but the Terrestrial ray was impotent compared with the weapons of the Titanians, and Stevens, snapping off the beam with a bitter imprecation, shot the visiray out toward the bare, black cone of the extinct volcano and studied it with care.

"Barkovis, I've got a thought!" he snapped into the microphone. "Their stronghold is in that mountain, and there's millions of them in there yet, coming out along their tunnels. They've got all the vegetation eaten away for miles, so there's nothing much left there to spread a fire if I go to work on that hill, and, I'll probably melt enough water to put out most of the fires

I start. Detail me a couple of ships to drop your fire-foam bombs on any little blazes that may spread, and I'll give them so much to worry about at home, that they'll forget all about Titania."

The *Forlorn Hope* darted toward the crater, followed closely by two of the dazzling globes. They circled the mountain until Stevens found a favorable point of attack—a stupendous vertical cliff of mingled rock and crystal, upon the base of which he trained his terrific infra-red projector.

"I'm going to draw a lot of power," he warned the Titanians then. "I'm giving this gun everything she'll take."

He drove the massive switches in, and as that dull red beam struck the cliff's base there was made evident the awful effect of a concentrated beam of real and pure heat upon such an utterly frigid world. Vast columns of fire roared aloft, helping Stevens, melting and destroying the very ground as the bodies of the Sedlor in that gigantic ant-heap burst into flames. Clouds of superheated steam roared upward, condensing into a hot rain which descended in destructive torrents upon the fastnesses of the centipedes. As the raging beam ate deeper and deeper into the base of the cliff, the mountain itself began to disintegrate; block after gigantic block breaking off and crashing down into the flaming, boiling, seething cauldron which was the apex of that ravening beam.

Hour after hour Stevens drove his intolerable weapon into the great mountain, teeming with Sedlorean life; and hour after hour a group of Titanian spheres stood by, deluging the surrounding plain with a flood of heavy fumes, through which the holocaust could not spread for lack of oxygen. Not until the mountain was gone—not until in its stead there lay a furiously boiling lake, its flaming surface hundreds of feet below the level of the plain—did Stevens open his power circuits and point the deformed prow of the *Forlorn Hope* toward Titania.

1 The moon subtends an angle of about one-half of a degree.

CHAPTER VII
The Return to Ganymede

"MUST you you go back to Ganymede?" Barkovis asked, slowly and thoughtfully. He was sitting upon a crystal bench beside the fountain, talking with Stevens, who, dressed in his bulging space-suit, stood near an airlock of the *Forlorn Hope*. "It seems a shame that you should face again those unknown, monstrous creatures who so inexcusably attacked us both without provocation."

"I'm not so keen on it myself, but I can't see any other way out of it," the Terrestrial replied. "We left a lot of our equipment there, you know; and even if I should build duplicates here, it wouldn't do us any good. These ten-nineteens are the most powerful transmitting tubes known when we left Tellus, but even their fields, dense as they are, can't hold an ultra-beam together much farther than about six astronomical units. So you see we can't possibly reach our friends from here with this tube; and your system of beam transmission won't hold anything together even that far, and won't work on any wave shorter than Roeser's Rays. We may run into some more of those little spheres, though, and I don't like the prospect. I wonder if we couldn't plate a layer of that mirror of yours upon the *Hope* and carry along a few of those bombs? By the way, what is that explosive—or is it something beyond Tellurian chemistry?"

"Its structure should be clear to you, although you probably could not prepare it upon Tellus because of your high temperature. It is nothing but nitrogen—twenty-six atoms of nitrogen combined to form one molecule of what you would call—N-twenty-six?"

"Wow!" Stevens whistled. "Crystalline, pentavalent nitrogen—no wonder it's violent!"

"We could, of course, cover your vessel with the mirror, but I am afraid that it would prove of little value. The plates are so hot that it would soon volatilize."

"Not necessarily," argued Stevens. "We could live in number one life-boat, and shut off the heat everywhere else. The life-boats are insulated from the structure proper, and the inner and outer walls of the structure are insulated from each other. With only the headquarters lifeboat warm, the outer wall could be held pretty close to zero absolute."

"That is true. The bombs, of course, are controlled by radio, and therefore may be attached to the outer wall of your vessel. We shall be glad to do these small things for you."

The heaters of the *Forlorn Hope* were shut off, and as soon as the outer shell had cooled to Titanian temperature, a corps of mechanics set to work. A machine very like a concrete mixer was rolled up beside the steel vessel, and into its capacious maw were dumped boxes and barrels of dry ingredients and many cans of sparkling liquid. The resultant paste was pumped upon the steel plating in a sluggish, viscid stream, which spread out into a thick and uniform coating beneath the flying rollers of the skilled Titanian workmen. As it hardened, the paste smoothed magically into the perfect mirror which covered the space-vessels of the satellite; and a full dozen of the mirror explosive bombs of this strange people were hung in the racks already provided.

"Once again I must caution you concerning those torpedoes," Barkovis warned Stevens. "If you use them all, very well, but do not try to take even one of them into any region where it is very hot, for it will explode and demolish your vessel. If you do not use them, destroy them before you descend into the hot atmosphere of Ganymede. The mirror will volatilize harmlessly at the temperature of melting mercury, but the torpedoes must be destroyed. Once more, Tellurians, we thank you for what you have done, and wish you well."

"Thanks a lot for *your* help—we still owe you something," replied Stevens. "If either of your power-plants go sour on you again, or if you need any more built, be sure to let us know—you can come close enough to the inner planets now on your own beam to talk to us on the ultra-communicator. We'll be glad to help you any way we can—and we may call on you for help again. Goodbye, Barkovis—goodbye, all Titania!"

He made his way through the bitterly cold shop into the control-room of their lifeboat, and while he was divesting himself of his heavy suit, Nadia lifted the *Forlorn Hope* into the blue-green sky of Titan, accompanied by an escort of the mirrored globes. Well clear of the atmosphere of the satellite, the terrestrial cruiser shot forward at normal acceleration, while the Titanian vessels halted and wove a pattern of blue and golden rays in salute to the departing guests.

"Well, Nadia, we're off—on a long trek, too."

"Said Wun Long Hop, the Chinese pee-lo," Nadia agreed. "Sure everything's all x, big boy?"

"To nineteen decimals," he declared. "You couldn't squeeze another frank into our accumulators with a proof-bar, and since they're sending us all the power we want to draw, we won't need to touch our batteries or tap our own beam until we're almost to Jupiter. To cap the climax, what it takes to make big medicine on those spherical friends of ours, we've got. We're not sitting on top of the world, ace—we've perched exactly at the apex of the entire universe!"

"How long is it going to take?"

"Don't know. Haven't figured it yet, but it'll be *beaucoup* days," and the two wanderers from far-distant Earth settled down to the routine of a long and uneventful journey.

They gave Saturn and his spectacular rings a wide berth and sped on, with ever-increasing velocity. Past the outer satellites, on and on, the good ship *Forlorn Hope* flew into the black-and-brilliant depths of interplanetary space. Saturn was an ever-diminishing disk beneath them: above them was Jupiter's thin crescent, growing ever larger and more bright, and the Monarch of the Solar System, remaining almost stationary day after day, increasing steadily in apparent diameter and in brilliance.

ALTHOUGH the voyage from Titan to Ganymede was long, it was not monotonous, for there was much work to be done in the designing and fabrication of the various units which were to comprise the ultra-radio transmitting station. In the various compartments of the *Forlorn Hope* there were sundry small motors, blowers, coils, condensers, force-field generators, and other items which Stevens could use with little or no alteration; but for the most part he had to build everything himself. Thus it was that time passed quickly; so quickly that Jupiter loomed large and the Saturnian beam of power began to attenuate almost before the Terrestrials realized that their journey was drawing to an end.

"Our beam's falling apart fast," Stevens read his meters carefully, then swung his communicator beam toward Jupiter. "We aren't getting quite enough power to hold our acceleration at normal—think I'll cut now, while we're still drawing enough to let the Titanians know we're off their beam. We've got lots of power of our own now; and we're getting pretty close to enemy territory, so they may locate that heavy beam. Have you found Ganymede yet?"

"Yes, it will be on the other side of Jupiter by the time we get there. Shall I detour, or put on a little more negative and wait for it to come around to this side?"

"Better wait, I think. The farther away we stay from Jupiter and the major satellites, the better."

"All x—it's on. Suppose we'd better start standing watches, in case some of them show up?"

"No use," he dissented. "I've been afraid to put out our electromagnetic detectors, as they could surely trace them in use. Without them, we couldn't spot an enemy ship even if we were looking right at it, except by accident; since they won't be lighted up and it's awfully hard to see anything out here, anyway. We probably won't know they're within a million kilometers until they put a beam on us. Barkovis says that this mirror will reflect any beam they can use, and I've already got a set of photo-cells in circuit to ring an alarm at the first flash off of our mirror plating. I'd like to get in the first licks myself, but I haven't been able to dope out any way of doing it. So you might as well sleep in your own room, as usual, and I'll camp here right under the panel until we get to Ganymede. There's a couple of little things I just thought of, though, that may help some; and I'm going to do 'em right now."

Putting on his space-suit, he picked up a power drill and went out into the bitter cold of the outer structure. There he attacked the inner wall of their vessel, and the carefully established inter-wall vacuum disappeared in a screaming hiss of air as the tempered point bit through plate after plate.

"What's the idea, Steve?" Nadia asked, when he had re-entered the control room. "Now you'll have all that pumping to do over again."

"Protection for the mirrors," he explained. "You see, they aren't perfect reflectors. There's a little absorption, so that some stuff comes through. Not much, of course; but enough to kill some of those Titanians and almost enough to ruin their ship got through in about ten minutes, and only one enemy was dealing it out. We can stand more than they could, of course, but the mirror itself won't stand much more heat than it was absorbing then. But with air in those spaces instead of vacuum, and with the whole mass of the *Hope*, except this one lifeboat, as cold as it is, I figure that there'll be enough conduction and convection through them to keep the outer wall and the mirror cold—cool enough, at least, to hold the mirror on for an hour. If only one ship tackles us, it won't be bad—but I figure that if there's only one, we're lucky."

STEVENS' fears were only too well grounded, for during the "evening" of the following day, while he was carefully scanning the heavens for some sign of enemy craft, the alarm bell over his head burst into its brazen clamor. Instantly he shot out the detectors and ultra-lights and saw not one, but six of the deadly globes—almost upon them, at point-blank range! One was already playing a beam of force upon the *Forlorn Hope*, and the other five went into action immediately upon feeling the detector impulses and perceiving that the weapon of their sister ship had encountered an unusual resistance in the material of that peculiarly mirrored wedge. As those terrific forces struck her, the terrestrial cruiser became a vast pyrotechnic set piece, a dazzling fountain of coruscant brilliance: for the mirror held. The enemy beams shot back upon themselves and rebounded in all directions, in the same spectacular exhibition of frenzied incandescence which had marked the resistance of the Titanian sphere to a similar attack.

But Stevens was not idle. In the instant of launching his detectors, as fast as he could work the trips, four of the frightful nitrogen bombs of Titan—all that he could handle at once—shot out into space, their rocket-tubes flaring viciously. The enemy detectors of course located the flying torpedoes immediately, but, contemptuous of material projectiles, the spheres made no attempt to dodge, but merely lashed out upon them with their ravening rays. So close was the range that they had no time to avoid the radio-directed bombs after discovering that their beams were useless against the unknown protective covering of those mirrored shells. There were four practically simultaneous detonations—silent, but terrific explosions as the pent-up internal energy of solid pentavalent nitrogen was instantaneously released—and the four insensately murderous spheres disappeared into jagged fragments of wreckage, flying wildly away from the centers of explosion. One great mass of riven and twisted metal was blown directly upon the fifth globe, and Nadia stared in horrified fascination at the silent crash as the entire side of the ship crumpled inward like a shell of cardboard under the awful impact. That vessel was probably out of action, but Stevens was taking no chances. As soon as he had clamped a pale blue tractor rod upon the sixth and last of the enemy fleet, he drove a torpedo through the gaping wall and into the interior of the helpless war-vessel. There he exploded it, and the awful charge, detonated in that confined space, literally tore the globular space-ship to bits.

"We'll show these jaspers what kind of trees make shingles!" he gritted between clenched teeth; and his eyes, hard now as gray iron, fairly emitted sparks as he launched four torpedoes upon the sole remaining globe of the squadron of the void. "I've had a lot of curiosity to know just what kind of unnatural monstrosities can possibly have such fiendish dispositions as they've got—but beasts, men or devils, they'll find they've grabbed something this time they can't let go of," and fierce blasts of energy ripped from the exhausts as he drove his missiles, at their highest possible acceleration, toward the captive sphere so savagely struggling at the extremity of his tractor beam.

But that one remaining vessel was to prove no such easy victim as had its sister ships. Being six to one, and supposedly invincible, the squadron had been overconfident and had attacked carelessly, with only its crippling slicing beams instead of its more deadly weapons of total destruction; and so fierce and hard had been Stevens' counter-attack that five of its numbers had been destroyed before they realized what powerful armament was mounted by that apparently crude, helpless, and innocuous wedge. The sixth, however, was fully warned, and every resource at the command of its hellish crew was now being directed against the *Forlorn Hope*.

Sheets, cones, and gigantic rods of force flashed and crackled. Space was filled with silent, devastating tongues of flame. The *Forlorn Hope* was dragged about erratically as the sphere tried to dodge those hurtling torpedoes; tried to break away from the hawser of energy anchoring her so solidly to her opponent. But the linkage held, and closer and closer Stevens drove the fourfold menace of his frightful dirigible bombs. Pressor beams beat upon them in vain. Hard driven as those pushers were, they could find no footing, but were reflected at many angles by that untouchable mirror and their utmost force scarcely impeded the progress of the rocket-propelled missiles. Comparatively small as the projectiles were, however, they soon felt the effects of the prodigious beams of heat enveloping them, and torpedo after torpedo exploded harmlessly in space as their mirrors warmed up and volatilized. But for each bomb that was lost, Stevens launched another, and each one came closer to its objective than had its predecessor.

Made desperate by the failure of his every beam, the enemy commander thought to use material projectiles himself—weapons abandoned long since by his race as antiquated and inefficient, but a few of which were still carried by the older types of vessels. One such shell was found and launched—but in the instant of its launching Stevens' foremost bomb struck its mark and exploded. So close were the other three bombs, that they also let go at the shock; and the warlike sphere, hemmed in by four centers of explosions, flew apart—literally pulverized. Its projectile, so barely discharged, did not explode—it was loaded with material which could be detonated only by the warhead upon impact or by a radio signal. It was, however, deflected markedly from its course by the force of the blast, so that instead of striking the *Forlorn Hope* in direct central impact, its head merely touched the apex of the mirror-plated wedge. That touch was enough. There was another appalling concussion, another blinding glare, and the entire front quarter of the terrestrial vessel had gone to join the shattered globes.

Between the point of explosion and the lifeboats there had been many channels of insulation, many bulkheads, many air-breaks, and compartment after compartment of accumulator cells. These had borne the brunt of the explosion, so that the control room was unharmed, and Stevens swung his communicator rapidly through the damaged portions of the vessels.

"How badly are we hurt, Steve—can we make it to Ganymede?"

Nadia was quietly staring over his shoulder into the plate, studying with him the pictures of destruction there portrayed as he flashed the projector from compartment to compartment.

"We're hurt—no fooling—but it might have been a lot worse," he replied, as he completed the survey. "We've lost about all of our accumulators, but we can land on our own beam, and landing power is all we want, I think. You see, we're drifting straight for where Ganymede will be, and we'd better cut out every bit of power we're using, even the heaters, until we get there. This lifeboat will hold heat for quite a while, and I'd rather get pretty cold than meet any more of that gang. I figured eight hours just before they met us, and we were just about drifting then. I think it is safe to say seven hours blind."

"But can't they detect us anyway? They may have sent out a call, you know."

"If we aren't using any power for anything, their electromagnetics are the only things we'll register on, and they're mighty short-range finders. Even if they should get that close to us, they'll probably think we're meteoric, since we'll be dead to their other instruments. Luckily we've got lots of air, so the chemical purifiers can handle it without power. I'll shut off everything and we'll drift it. Couldn't do much of anything, anyway—even our shop out there won't hold air. But we can have light. We've got acetylene emergency lamps, you know, and we don't need to economize on oxygen."

"Perhaps we'd better run in the dark. Remember what you told me about their possible visirays, and that you've got only two bombs left."

"All x; that would be better. If I forget it, remind me to blow up those before we hit the atmosphere of Ganymede, will you?" He opened all the power switches, and, every source of ethereal vibrations cut off, the *Forlorn Hope* drifted slowly on, now appearing forlorn indeed.

SEVEN hours dragged past: seven age-long hours during which the two sat tense, expecting they knew not what, talking only at intervals and in subdued tones. Stevens then snapped on the communicator beam just long enough to take an observation upon Ganymede. Several such brief glimpses were taken; then, after a warning word to his companion, he sent out and exploded the nitrogen bombs. He then threw on the power, and the vessel leaped toward the satellite under full acceleration. Close to the atmosphere it slanted downward in a screaming, fifteen-hundred-mile drive; and soon the mangled wedge dropped down into the little canyon, which for so long had been "home."

"Well, colonel, home again!" Stevens exulted as he neutralized the controls. "There's that falls, our power plant, the catapults, 'n' everything. Now, unless something interrupts us again; we'll run up our radio tower and give Brandon the long yell."

"How much more have you got to do before you can start sending?"

"Not an awful lot. Everything built—all I've got to do is assemble it. I should be able to do it easily in a week. Hope nothing else happens—if I drag you into any more such messes as those we've just been getting out of by the skin of our teeth, I'll begin to wish that we had started out at first to drift it back to Tellus in the *Hope.* Let's see how much time we've got. We should start shooting one day after an eclipse, so that we'll have five days to send. You see, we don't want to point our beam too close to Jupiter or to any of the large satellites, because the enemy might live there and might intercept it. We had an eclipse yesterday—so one week from today, at sunrise, I start shooting."

"But Earth's an evening star now; you can't see it in the morning."

"I'm not going to aim at Tellus. I'm shooting at Brandon, and he's never there for more than a week or two at a stretch. They're prowling around out in space somewhere almost all the time."

"Then how can you possibly hope to hit them?"

"It may be quite a job of hunting, but not as bad as you might think. They probably aren't much, if any, outside the orbit of Mars, and they usually stay within a couple of million kilometers or so of the Ecliptic, so we'll start at the sun and shoot our beam in a spiral to cover that field. We ought to be able to hit them inside of twelve hours, but if we don't, we'll widen our spiral and keep on trying until we do hit them."

"Heavens, Steve! Are you planning on telegraphing steadily for days at a time?"

"Sure, but not by hand, of course—I'll have an automatic sender and automatic pointers."

Stevens had at his command a very complete machine-shop, he had an ample supply of power, and all that remained for him to do was to assemble the parts which he had built during the long journey from Titan to Ganymede. Therefore, at sunrise of the designated day, he was ready, and, with Nadia hanging breathless over his shoulder, he closed the switch, a toothed wheel engaged a delicate interrupter, and a light sounder began its strident chatter.

"Ganymede point oh four seven ganymede point oh four seven ganymede point oh four seven..." endlessly the message was poured out into the ether, carried by a tight beam of ultra-vibrations and driven by forces sufficient to propel it well beyond the opposite limits of the orbit of Mars.

"What does it say? I can't read code."

Stevens translated the brief message, but Nadia remained unimpressed.

"But it doesn't say anything!" she protested. "It isn't addressed to anybody, it isn't signed—it doesn't tell anybody anything about anything."

"It's all there, ace. You see, since the beam is moving sidewise very rapidly at that range and we're shooting at a small target, the message has to be very short or they won't get it all while the beam's on 'em—it isn't as though we were broadcasting. It doesn't need any address, because nobody but the *Sirius* can receive it—except possibly the Jovians. They'll know who's sending it without any signature. It tells them that Ganymede wants to receive a message on the ultra-band centering on forty-seven thousandths. Isn't that enough?"

"Maybe. But suppose some of them live right here on Ganymede—you'll be shooting right through the ground all night—or suppose that even if they don't live here, that they can find our beam some way? Or suppose that Brandon hasn't got his machine built yet, or suppose that it isn't turned on when our beam passes them, or suppose they're asleep then? A lot of things might happen."

"Not so many, ace—your first objection is the only one that hasn't got more holes in it than a sieve, so I'll take it first. Since our beam is only a meter in diameter here and doesn't spread much in the first few million kilometers, the chance of direct reception by the enemy, even if they do live here on Ganymede, is infinitesimally small. But I don't believe that they live here—at least, they certainly didn't land on this satellite. As you suggest, however, it is conceivable that they may have detector screens delicate enough to locate our beam at a distance; but since in all probability that means a distance of hundreds of thousands of kilometers, I think it highly improbable. We've got to take the same risk anyway, no matter what we do, whenever we start to use any kind of driving power, so there's no use worrying about it. As for your last two objections, I know Brandon and I know Westfall. Brandon will have receivers built that will take in any wave possible of propagation, and Westfall, the cautious old egg, will have them running twenty-four hours a day, with automatic recorders, finders, and everything else that Brandon can invent—and believe me, sweetheart, that's a lot of stuff!"

"It's wonderful, the way you three men are," replied Nadia thoughtfully, reading between the lines of Stevens' utterance. "They knew that you were on the *Arcturus*, of course—and they knew that if you were alive you'd manage in some way to get in touch with them. And you, away out here after all this time, are superbly confident that they are expecting a call from you. That, I think, is one of the finest things I ever heard of."

"They're two of the world's best—absolutely." Nadia looked at him, surprised, for he had not seen anything complimentary to himself in her remark. "Wait until you meet them. They're men, Nadia—real men. And speaking of meeting them—please try to keep on loving me after you meet Norm Brandon, will you?"

"Don't be a simp!" her brown eyes met his steadily. "You didn't mean that—you didn't even say it, did you?"

"Back it comes, sweetheart! But knowing myself and knowing those two...."

"Stop it! If Norman Brandon or Quincy Westfall had been here instead of you, or both of them together, we'd have been here from now on—we wouldn't even have gotten away from the Jovians!"

"NOW it's your turn to back water, guy!"

"Well, maybe, a little—if both of them were here, they ought to equal you in some things. Brandon says himself that he and Westfall together make one scientist—Dad says he says so."

"You don't want to believe everything you hear. Neither of them will admit that he knows anything or can do anything—that's the way they are."

"Dad has told me a lot about them—how they've always been together ever since their undergraduate days. How they studied together all over the world, even after they'd been given all the degrees loose. How they even went to the other planets to study—to Mars, where they had to live in space-suits all the time, and to Venus, where they had to take ultra-violet treatments every day to keep alive. How they learned everything that everybody else knew and then went out into space to find out things that nobody else ever dreamed of. How you came to join them, and what you three have done since. They're fine, of course—but they aren't *you*," she concluded passionately.

"No, thank Heaven! I know you love me, Nadia, just as I love you—you know I never doubted it. But you'll like them, really. They're a wonderful team. Brandon's a big brute, you know—fully five centimeters taller than I am, and he weighs close to a hundred kilograms—and no lard, either. He's wild, impetuous, always jumping at conclusions and working out theories that seem absolutely ridiculous, but they're usually sound, even though impractical. Westfall's the practical member—he makes Norm pipe down, pins him down to facts, and makes it possible to put his hunches and wild flashes of genius into workable form. Quince is a...."

"Now *you* pipe down! I've heard you rave *so* much about those two—I'd lots rather rave about you, and with more reason. I wish that sounder would start sounding."

"Our first message hasn't gone half way yet. It takes about forty minutes for the impulse to get to where I think they are, so that even if they got the first one and answered it instantly, it would be eighty minutes before we'd get it. I sort of expect an answer late tonight, but I won't be disappointed if it takes a week to locate them."

"I will!" declared the girl, and indeed, very little work was done that day by either of the castaways.

Slowly the day wore on, and the receiving sounder remained silent. Supper was eaten as the sun dropped low and disappeared, but they felt no desire to sleep. Instead, they went out in front of the steel wall, where Stevens built a small campfire. Leaning back against the wall of their vessel, they fell into companionable silence, which was suddenly broken by Stevens.

"Nadia, I just had a thought. I'll bet four dollars I've wasted a lot of time. They'll certainly have automatic relays on Tellus, to save me the trouble of hunting for them, but like an idiot I never thought of it until just this minute, in spite of the speech I made you about them. I'm going to change those directors right now."

"That's quite a job, isn't it?"

"No, only a few minutes."

"Do it in the morning; you've done enough for one day—maybe you've hit them already, any way."

They again became silent, watching Jupiter, an enormous moon some seven degrees in apparent diameter.

"Steve, I simply can't get used to such a prodigious moon! Look at the stripes, and look at that perfectly incredible...."

A gong sounded and they both jumped to their feet and raced madly into the *Hope*. The ultra-receiver had come to life and the sounder was chattering insanely—someone was sending with terrific speed, but with perfect definition and spacing.

"That's Brandon's fist—I'd know his style anywhere," Stevens shouted, as he seized notebook and pencil.

"Tell me what it says, quick, Steve!" Nadia implored.

"Can't talk—read it!" Stevens snapped. His hand was flying over the paper, racing to keep up with the screaming sounder.

"...ymede all x stevens ganymede all x stevens ganymede all x placing and will keep sirius on plane between you and tellus circle fifteen forty north going tellus first send full data spreading beam to cover circle fifteen forty quince suggests possibility this message intercepted and translated personally I think such translation impossible and that he is wilder than a hawk but just in case they should be supernaturally intelligent...."

Stevens stopped abruptly and stared at the vociferous sounder.

"Don't stop to listen—keep on writing!" commanded Nadia.

"Can't," replied the puzzled mathematician. "It doesn't make sense. It sounds intelligent—it's made up of real symbols of some kind or other, but they don't mean a thing to me."

"Oh, I see—he's sending mush on purpose. Read the last phrase!"

"Oh, sure—'mush' is right," and with no perceptible break the signals again became intelligible.

"... if they can translate that they are better scholars than we are signing off until hear from you brandon."

THE sounder died abruptly into silence and Nadia sobbed convulsively as she threw herself into Stevens' arms. The long strain over, the terrible uncertainty at last dispelled, they were both incoherent for a minute—Nadia glorifying the exploits of her lover, Stevens crediting the girl herself and his two fellow-scientists with whatever success had been achieved. A measure of self-control regained, Stevens cut off his automatic sender, changed the adjustments of his directors and cut in his manually operated sending key.

"What waves are you using, anyway?" asked Nadia, curiously. "They must be even more penetrating than Roeser's Rays, to have such a range, and Roeser's Rays go right through a planet without even slowing up."

"They're of the same order as Roeser's—that is, they're sub-electronic waves of the fourth order—but they're very much shorter, and hence more penetrating. In fact, they're the shortest waves yet known, so short that Roeser never even suspected their existence."

"Suppose there's a Jovian space-ship out there somewhere that intercepts our beams. Couldn't they locate us from it?"

"Maybe, and maybe not—we'll just have to take a chance on that. That goes right back to what we were talking about this morning. They might be anywhere, so the chance of hitting one is very small. It isn't like hitting the *Sirius,* because we knew within pretty narrow limits where to look for her, and even at that we had to hunt for her for half a day before we hit her.

We're probably safe, but even if they should have located us, we'll probably be able to hide somewhere until the *Sirius* gets here. Well, the quicker I get busy sending the dope, the sooner they can get started."

"Tell them to be sure and bring me all my clothes they can find, a gallon of perfume, a barrel of powder, and a carload of Delray's Fantasie chocolates—I've been a savage so long that I want to wallow in luxury for a while."

"I'll do that—and I want some real cigarettes!"

Stevens first sent a terse, but complete account of everything that had happened to the *Arcturus*, and a brief summary of what he and Nadia had done since the cutting up of the IPV. The narrative finished, he launched into a prolonged and detailed scientific discussion of the enemy and their offensive and defensive weapons. He dwelt precisely and at length upon the functioning of everything he had seen. Though during the long months of their isolation he had been too busy to do any actual work upon the weapons of the supposed Jovians, yet his keen mind had evolved many mathematical and physical deductions, hypotheses, and theories, and these he sent out to the *Sirius*, concluding:

"There's all the dope I can give you. Figure it out, and don't come at all until you can come loaded for bear; they're bad medicine. Call us occasionally, to keep us informed as to when to expect you, but don't call too often. We don't want them locating you, and if they should locate us through your ray or ours, it would be just too bad. So-long. Stevens and Newton."

Nadia had insisted upon staying up and had been brewing pot after pot of her substitutes for coffee while he sat at the key; and it was almost daylight when he finally shut off the power and arose, his right arm practically paralyzed from the unaccustomed strain of hours of telegraphing.

"Well, sweetheart, that's that!" he exclaimed in relief. "Brandon and Westfall are on the job. Nothing to do now but wait, and study up on our own account on those Jovians' rays. This has been one long day for us, though, little ace, and I suggest that we sleep for about a week!"

CHAPTER VIII
Callisto to the Rescue

ALL humanity of Callisto, the fourth major satellite of Jupiter, had for many years been waging a desperate and apparently hopeless defense against invading hordes of six-limbed beings. Every city and town had long since been reduced to level fields of lava by the rays of the invaders. Every building and every trace of human civilization had long since disappeared from the surface of the satellite. Far below the surface lay the city of Zbardk, the largest of the few remaining strongholds of the human race. At one portal of the city a torpedo-shaped, stubby-winged rocket plane rested in the carriage of a catapult. Near it the captain addressed briefly the six men normally composing his crew.

"Men, you already know that our cruise today is not an ordinary patrol. We are to go to One, there to destroy a base of the hexans. We have perhaps one chance in ten thousand of returning. Therefore I am taking only one man—barely enough to operate the plane. Volunteers step one pace forward."

The six stepped forward as one man, and a smile came over the worn face of their leader as he watched them draw lots for the privilege of accompanying him to probable death. The two men entered the body of the torpedo, sealed the openings and waited.

"Free exits?" snapped the Captain of the Portal, and twelve keen-eyed observers studied minutely screens and instrument panels connected to the powerful automatic lookout stations beneath the rims of the widely separated volcanic craters from which their craft could issue into Callisto's somber night.

"No hexan radiation can be detected from Exit Eight," came the report. The Captain of the Portal raised an arm in warning, threw in the guides, and the two passengers were hurled violently backward, deep into their cushioned seats, as the catapult shot their plane down the runway. As the catapult's force was spent automatic trips upon the undercarriage actuated the propelling rockets and mile after mile, with rapidly mounting velocity, the plane sped through the tube. As the exit was approached, the tunnel described a long vertical curve, so that when the opening into the shaft of the crater was reached and the undercarriage was automatically detached, the vessel was projected almost vertically upward. Such was its velocity and so powerful was the liquid propellant of its rocket motors, that the eye could not follow the flight of the warship as it tore through the thin layer of the atmosphere and hurled itself out into the depths of space.

"Did we get away?" asked the captain, hands upon his controls and eyes upon his moving chart of space.

"I believe so, sir," answered the other officer, at the screens of the six periscopic devices which covered the full sphere of vision. "No reports from the rim, and all screens blank." Once more a vessel had issued from the jealously secret city of Zbardk without betraying its existence to the hated and feared hexans.

For a time the terrific rocket motors continued the deafening roar of their continuous explosions, then, the desired velocity having been attained, they were cut out and for hours the good ship "Bzark" hurtled on through the void at an enormous but constant speed toward the distant world of One, which it was destined never to reach.

"Captain Czuv! Hexan radiation, coordinates twenty two, fourteen, area six!" cried the observer, and the commander swung his own telescopic finder into the indicated region. His hands played over course and distance plotters for a brief minute, and he stared at his results in astonishment.

"I never heard of a hexan traveling that way before," he frowned. "Constant negative acceleration and in a straight line. He must think that we have been cleared out of the ether. Almost parallel to us and not much faster—even at this long range, it is an easy kill unless he starts dodging, as usual."

As he spoke, he snapped a switch and from a port under the starboard wing there shot out into space a small package of concentrated destruction—a rocket-propelled, radio-controlled torpedo. The rockets of the tiny missile were flaming, but that flame was visible only from the rear and no radio beam was upon it. Czuv had given it precisely the direction and acceleration necessary to make it meet the hexan sphere in central impact, provided that sphere maintained its course and acceleration unchanged.

"Shall I direct the torpedo in the case the hexan shifts?" asked the officer.

"I think not. They can, of course, detect any wave at almost any distance, and at the first sign of radioactivity they would locate and destroy the bomb. They also, in all probability, would destroy us. I would not hesitate to attack them on that account alone, but we must remember that we are upon a more important mission than attacking one hexan ship. We are far out of range of their electromagnetic detectors, and our torpedo will have such a velocity that they will have no time to protect themselves against it after detection. Unless they shift in the next few seconds, they are lost. This is the most perfect shot I ever had at one of them, but one shot is all I dare risk—we must not betray ourselves."

COURSE, lookout, and rank forgotten, the little crew of two stared into the narrow field of vision, set at its maximum magnification. The instruments showed that the enemy vessel was staying upon its original course. Very soon the torpedo came within range of the detectors of the hexans. But as Captain Czuv had foretold, the detection was a fraction of a second too late, rapidly as their screens responded, and the two men of Zbardk uttered together a short, fierce cry of joy as a brilliant flash of light announced the annihilation of the hexan vessel.

"But hold!" The observer stared into his screen. "Upon that same line, but now at constant velocity, there is still a very faint radiation, of a pattern I have never seen before."

"I think ... I believe ..." the captain was studying the pattern, puzzled. "It must be low frequency, low-tension electricity, which is never used, so far as I know. It may be some new engine of destruction, which the hexan was towing at such a distance that the explosion of our torpedo did not destroy it. Since there are no signs of hexan activity and since it will not take much fuel, we shall investigate that radiation."

Tail and port-side rockets burst into roaring activity and soon the plane was cautiously approaching the mass of wreckage, which had been the IPV *Arcturus*.

"Human beings, although of some foreign species!" exclaimed the captain, as his vision-ray swept through the undamaged upper portion of the great liner and came to rest upon Captain King at his desk.

Although the upper ultra-lights of the Terrestrial vessel had been cut away by the hexan plane of force, jury lights had been rigged, and the two commanders were soon trying to communicate with each other. Intelligible conversation was, of course, impossible, but King soon realized that the visitors were not enemies. At their pantomimed suggestion he put on a space-suit and wafted himself over to the airlock of the Callistonian warplane. Inside the central compartment, the strangers placed over his helmet a heavily wired harness, and he found himself instantly in full mental communication with the Callistonian commander. For several minutes they stood silent, exchanging thoughts with a rapidity impossible in any language; then, dressed in space-suits, both leaped lightly across the narrow gap into the still open outer lock of the terrestrial liner. King watched Czuv narrowly after the pressure began to collapse his suit, but the stranger made no sign of distress. He had been right in his assurance that the extra pressure would scarcely inconvenience him. King tore off his helmet, issued a brief order, and soon every speaker in the *Arcturus* announced:

"All passengers and all members of the crew except lookouts on duty will assemble immediately in Saloon Three to discuss a possible immediate rescue."

The subject being one of paramount interest, it was a matter of minutes until the full complement of two hundred men and women were in the main saloon, clinging to hastily rigged hand lines, closely packed before the raised platform upon which were King and Czuv, wired together with the peculiar Callistonian harness. To most of the passengers, familiar with the humanity of three planets, the appearance of the stranger brought no surprise; but many of them stared in undisguised amazement at his childish body, his pale, almost colorless skin, his small, weak legs and arms, and his massive head.

"Ladies and gentlemen!" Captain King opened the meeting. "I introduce to you Captain Czuv, of the scout cruiser *Bzarvk*, of the only human race now living upon the fourth large satellite of Jupiter, which satellite we know as Callisto. I am avoiding their own names as much as possible, because they are almost unpronounceable in English or Interplanetarian. This device that you see connecting us is a Callistonian thought transformer, by means of which any two intelligent beings can converse without language. Our situation is peculiar, and in order that you may understand fully what lies ahead of us, the captain will now speak to you, through me—that is, what follows will be spoken by Captain Czuv, of the *Bzarvk*, but he will be using my vocal organs."

"Friends from distant Tellus," King's voice went on, almost without a break, "I greet you. I am glad, for your sake as well as our own, that your vessel was able to destroy the hexan ship holding you captive, and whose crew would have killed you all as soon as they had landed your vessel and had read your minds. I regret bitterly that we can do so little for you, for only the representatives of a human civilization being exterminated by a race of highly intelligent monsters can fully realize how desirable it is for all the various races of humanity to assist and support each other. In order that you may understand the situation, it is necessary that I delve at some length into ancient history, but we have ample time. In about ..." he broke off, realizing that the two races had no thought in common in the measure of time.

"One-half time of rotation of Great Planet upon axis?" flashed from Czuv's brain, and "About five hours," King's mind flashed back.

"It will be about five hours before any steps can be taken, so that I feel justified in using a brief period for explanation. In the evolution of the various forms of life upon Callisto, two genera developed intelligence far ahead of all others. One genus was the human, as you and I; the other the hexan. This creature, happily unknown to you of the planets nearer our common sun, is the product of an entirely different evolution. It is a six-limbed animal, with a brain equal to our own—one perhaps in some ways superior to our own. They have nothing in common with humanity, however; they have few of our traits and fewer of our mental processes. Even we who have fought them so long can scarcely comprehend the chambers of horror that are their minds. Even were I able to paint a sufficiently vivid picture with words, you of Earth could not begin to understand their utter ruthlessness and inhumanity, even among themselves. You would believe that I was lying, or that my viewpoint was warped. I can say only that I hope most sincerely that none of you will ever get better acquainted with them."

"**A**GES ago, then, the human and the hexan developed upon all four of the major satellites of the Great Planet, which you know as Jupiter, and upon the north polar region of Jupiter itself. By what means the two races came into being upon worlds so widely separated in space we know not—we only know it to be the fact. Human life, however, could not long endure upon Jupiter. The various human races, after many attempts to meet conditions of life there by variations in type fell before the hexans; who, although very small in size upon the planet, thrived there amazingly. Upon the three outer satellites humanity triumphed, and many hundreds of cycles ago the hexans of those satellites were wiped out, save for an occasional tribe of savages of low intelligence who lived in various undesirable portions of the three worlds. For ages then there was peace upon Callisto. Here is the picture at that time—upon Jupiter the hexans; upon Io hexans and humans, waging a ceaseless and relentless war of mutual extermination; upon the three outer satellites humanity in undisturbed and unthreatened peace. Five worlds, each ignorant of life upon any other.

"As I have said, the hexans of Jupiter were, and are, diabolically intelligent. Driven probably by their desire to see what lay beyond their atmosphere of eternal cloud, to the penetration of which their eyesight was attuned, they developed the space-ship; and effected a safe landing, first upon the barren, airless moonlet nearest them, and then upon fruitful Io. There they made common cause with the hexans against the humans, and in space of time Ionian humanity ceased to exist. Much traffic and interbreeding followed between the hexans of Jupiter and those of Io, resulting in time in a race intermediate in size between the parent stocks and equally at home in the widely variant air pressures and gravities of planet and satellite. Soon their astronomical instruments revealed the cities of Europa to their gaze, and as soon as they discovered that the civilization of Europa was human, they destroyed it utterly, with the insatiable blood lust that is their heritage.

"In the meantime the human civilizations of Ganymede and Callisto had also developed instruments of power. Observing the cities upon the other satellites, many scientists studied intensively the problem of space navigation, and finally there was some commerce between the two outer satellites at favorable times. Finally, vessels were also sent to Io and to Europa, but none of them returned. Knowing then what to expect, Ganymede and Callisto joined forces and prepared for war. But our science, so long attuned to the arts of peace, had fallen behind lamentably in the devising of more and ever more deadly instruments of destruction. Ganymede fell, and in her fall we read our own doom. Abandoning our cities, we built anew underground. Profiting from lessons learned full bloodily upon Ganymede, we resolved to prolong the existence of the human race as long as possible.

"The hexans were, and are masters of the physical science. They command the spectrum in a way undreamed of. Their detectors reveal etheric disturbances at unbelievable distances, and they have at their beck and call forces of staggering magnitude. Therefore in our cities is no electricity save that which is wired, shielded, and grounded; no broadcast radio; no source whatever of etheric disturbances save light—and our walls are fields of force which we believe to be impenetrable to any searching frequency capable of being generated. Now I am able to picture to you the present.

"We are the last representatives of the human race in the Jovian planetary system. Our every trace upon the surface has been obliterated. We are hiding in our holes in the ground, coming out at night by stealth so that our burrows shall not be revealed to the hexans. We are fighting for time in which our scientists may learn the secrets of power—and fearing, each new day, that the enemy may have so perfected their systems of rays that they will be able to detect us and destroy us, even in our underground and heavily shielded retreats, by means of forces even more incomprehensible than those they are now employing.

"Therefore, friends, you see how little we are able to do for you, we a race fighting for our very existence and doomed to extinction save for a miracle. We cannot take you to Callisto, for it is besieged by the hexans and the driving forces of your lifeboats, practically broadcast as they are, would be detected and we should all be destroyed long before we could reach safety. Captain King and I have pondered long and have been able to see only one course of action. We are drifting at constant velocity, using no power, and with all save the most vitally necessary machinery at rest. Thus only may we hope to avoid detection during the next two hours.

"Our present course will take us very close to Europa, which the hexans believe to be like Ganymede, entirely devoid of civilized life. Its original humanity was totally destroyed, and all its civilized hexans are finding shelter from our torpedoes upon Jupiter until we of Callisto shall likewise have been annihilated. The temperature of Europa will suit you. Its atmosphere, while less dense than that to which you are accustomed, will adequately support your life. If we are not detected in the course of the next few hours we can probably land upon Europa in safety, since its neighborhood is guarded but loosely. In fact, we have a city there, as yet unsuspected by the hexans, in which our scientists will continue to labor after Callisto's civilization shall have disappeared. We think that it will be safe to use your power for the short time necessary to effect a landing. We shall land in a cavern, in a crater already in communication with our city. In that cavern, instructed and aided by some of us, you will build a rocket vessel—no rays can be used because of the hexans—in which you will be able to travel to a region close enough to your earth so that you can call for help. You will not be able to carry enough fuel to land there—in fact, nearly all the journey will have to be made without power, traveling freely in a highly elongated orbit around the sun—but if you escape the hexans, you should be able to reach home safely, in time. It is for the consideration of this plan that this meeting has been called."

"JUST one question," Breckenridge spoke. "The hexans are intelligent. Why are they leaving Europa and Ganymede so unguarded that human beings can move back there and that we can land there, all undetected?"

"I will answer that question myself," replied King. "Captain Czuv did not quite do justice to his own people. It is true that they are being conquered, but for every human life that is taken, a thousand hexans die, and for every human ship that is lost, twenty hexan vessels are annihilated in return. While the hexans are masters of rays, the humans are equally masters of explosives and of mechanisms. They can hit a perfect score upon any target in free space whose course and acceleration can be determined, at any range up to five thousand kilometers, and they have explosives thousands of times as powerful as any known to us. Ray screens are effective only against rays, and the hexans cannot destroy anything they cannot see before it strikes them. So it is that all the hexan vessels except those necessary to protect their own strongholds, are being concentrated against Callisto. They cannot spare vessels to guard uselessly the abandoned satellites. Because of the enormously high gravity of Jupiter the hexans there are safe from human attack save for ineffectual long-range bombardment, but Io is being attacked constantly and it is probable that in a few more years Io also will be an abandoned world. Some of you may have received the impressions that the hexans are to triumph immediately, but such an idea is wrong. The humans can, and will, hold out for a hundred years or more unless the enemy perfects a destructive ray of the type referred to. Even then, I think that our human cousins will hold out a long time. They are able men, fighters all, and their underground cities are beautifully protected."

There was little argument. Most of the auditors could understand that the suggested course was the best one possible. The remainder were so stunned by the unbelievable events of the attack that they had no initiative, but were willing to follow wherever the more valiant spirits led. It was decided that no attempt should be made to salvage any portion of the *Arcturus*, since any such attempt would be fraught with danger and since the wreckage would be of little value. The new vessel was to be rocket driven and was to be built of Callistonian alloys. Personal belongings were moved into lifeboats, doors were closed, and there ensued a painful period of waiting and suspense.

The stated hour was reached without event—no hexan scout had come close enough to them to detect the low-tension radiation of the vital machinery of the *Arcturus*, cut as it was to the irreducible minimum and quite effectively grounded as it was by the enormous mass of her shielding armor. At a signal from Captain Czuv the pilot of each lifeboat shot his tiny craft out into space and took his allotted place in the formation following closely behind the *Bzarvk*, flying toward Europa, now so large in the field of vision that she resembled more a world than a moon. Captain King, in the Callistonian vessel, transmitted to Breckenridge the route and flight data given him by the navigator of the winged craft. The chief pilot, flying "point," in turn relayed more detailed instructions to the less experienced pilots of the other lifeboats.

Soon the surface of Europa lay beneath them; a rugged, cratered, and torn topography of mighty ranges of volcanic mountains. Most of the craters were cold and lifeless; but here and there a plume of smoke and steam betrayed the presence of vast, quiescent forces. Straight down one of those gigantic lifeless shafts the fleet of space craft dropped—straight down a full two miles before the landing signal was given. At the bottom of the shaft a section of the rocky wall swung aside, revealing the yawning black mouth of a horizontal tunnel. At intervals upon its roof there winked into being almost invisible points of light. Along that line of lights the lifeboats felt their way, coming finally into a huge cavern, against one sheer metal wall of which they parked in an orderly row. Roll was called, and the terrestrials walked, as well as they could in the feeble gravity of the satellite, across the vast chamber and into a conveyance somewhat resembling a railway coach, which darted away as soon as the doors were shut. For hundreds of miles that strange tunnel extended, and as the car shot along door after door of natural rock opened before it, and closed as soon as it had sped through. In spite of the high velocity of the vehicle, it required almost two hours to complete the journey. Finally, however, it slowed to a halt and the Terrestrial visitors disembarked at a portal of the Europan city of the Callistonians.

"Attention!" barked Captain King. "The name of this city, as nearly as I can come to it in English, is *WRUZK*. 'Roosk' comes fairly close to it and is easier to pronounce. We must finish our trip in small cars, holding ten persons each. We shall assemble again in the building in which we have been assigned quarters. The driver of each car will lead his passengers to the council room in which we shall meet."

"Oh, what's the use—this is horrible, horrible—we might as well die!" a nervous woman shrieked, and fainted.

"Such a feeling is, perhaps, natural," King went on, after the woman had been revived and quiet had been restored, "but please control it as much as possible. We are alive and well, and will be able to return to Tellus eventually. Please remember that these people are putting themselves to much trouble and inconvenience to help us, desperate as their own situation is, and conduct yourselves accordingly."

The rebuke had its effect, and with no further protest the company boarded the small cars, which shot through an opening in the wall and into a street of that strange subterranean city. Breckenridge, in the last car to leave the portal, studied his surroundings with interest as his conveyance darted through the gateway. More or less a fatalist by nature and an adventurer, of course, since no other type existed among the older spacehounds of the IPC, he was intensely interested in every new phase of their experience, and was no whit dismayed or frightened.

HE found himself seated in a narrow canoe of metal, immediately behind the pilot, who sat at a small control panel in the bow. Propelled by electromagnetic fields above a single rail, upon lightly touching and noiseless wheels, the terrestrial pilot saw with keen appreciation the manner in which switch after switch ahead of them obeyed the impulses sent ahead from the speeding car. The streets were narrow and filled with monorails; pedestrians pursued their courses upon walks attached to the walls of the buildings, far above the level of the streets. The walls were themselves peculiar, rising as they did stark, unbroken, windowless expanses of metal, merging into and supporting a massive roof of the same silvery metal. Walls and roof alike reflected a soft, yet intense, white light. Soon a sliding switch ahead of them shot in and simultaneously an opening appeared in the blank metal wall of a building. Through the opening the street-car flew, and as the pilot slowed the canoe to a halt, the door slid smoothly shut behind them. Parking the car beside a row of its fellows, the Callistonian driver indicated that the Terrestrials were to follow him and led the way into a large hall. There the others from the *Arcturus* were assembled, facing Captain King, who was standing upon a table.

"Fellow travelers," King addressed them, "our course of action has been decided. There are two hundred three of us. There will be twenty sections of ten persons, each section being in charge of one of the officers of the *Arcturus*. Doctor Penfield, our surgeon, a man whose intelligence, fairness, and integrity are unquestioned, will be in supreme command. His power and authority will be absolute, limited only by the Callistonian Council. He will work in harmony with the engineer, who is to direct the entire project of building the new vessel. Each of you will be expected to do whatever he can—the work you will be asked to do will be well within your powers, and you will each have ample leisure for recreation, study, and amusement, of all of which you will find unsuspected stores in this underground community. You will each be registered and studied by physicians, surgeons, and psychologists; and each of you will have prescribed for him the exact diet that is necessary for his best development. You will find this diet somewhat monotonous, compared to our normal fare of natural products, since it is wholly synthetic; but that is one of the minor drawbacks that must be endured. Chief Pilot Breckenridge and I will not be with you. In some small and partial recompense for what they are doing for us all, he and I are going with Captain Czuv to Callisto, there to see whether or not we can aid them in any way in the fight against the hexans. One last word—Doctor Penfield's rulings will be the products of his own well-ordered mind after consultation and agreement with the Council of this city, and will be for the best good of all. I do not anticipate any refusal to cooperate with him. If, however, such refusal should occur, please remember that he is a despot with absolute power, and that anyone obstructing the program by refusing to follow his suggestions will spend the rest of his time here in confinement and will go back to Tellus in irons, if at all. In case Chief Pilot Breckenridge and I should not see you again, we bid you goodbye and wish you a safe voyage—but we expect to go back with you."

Brief farewells were said and captain and pilot accompanied Czuv to one of the little street-cars. Out of the building it dashed and down the crowded but noiseless thoroughfare to the portal. Signal lights flashed briefly there and they did not stop, but tore on through the portal and the tunnel, with increasing speed.

"Don't have to transfer to a big car, then?" asked Breckenridge.

"No," King made answer. "Small cars can travel these tubes as well as the large ones, and on much less power. In the city the wheels touch the rails lightly, not for support, but to make contacts through which traffic signals are sent and received. In the tunnels the wheels do not touch at all, as signaling is unnecessary—the tunnels being used infrequently and by but one vehicle at a time. No trolleys, tracks, or wires are visible, you notice. Everything is hidden from any possible visiray of the hexans."

"How about their power?"

"I don't understand it very well—hardly at all, in fact."

"It is quite simple." To the surprise of both Terrestrials, Czuv was speaking English, but with a strong and very peculiar accent; slighting all the vowels and accenting heavily the consonant sounds. "The car no longer requires my attention, so I am now free to converse. You are surprised at my knowing your language? You will speak mine after a few more applications of the thought exchanger. I am speaking with a vile accent, of course, but that is merely because my vocal organs are not accustomed to making vowel sounds. Our power is obtained by the combustion of gases in highly efficient turbines. It is transmitted and used as direct current, our generator and motors being so constructed that they can produce no etheric disturbances capable of penetrating the shielding walls of our city. The city was built close to deposits of coal, oil, and gas of sufficient amount to support our life for thousands of years; for from these deposits come power, food, clothing, and all the other necessities and luxuries of our lives. Strong fans draw air from various extinct craters, force it through ventilating ducts into every room and recess of the city, and exhaust it into the shaft of a quiescent volcano, in whose gaseous outflow any trace of our activities is, of course, imperceptible. For obvious reasons no rockets or combustion motors are used in the city proper."

THUS Captain Czuv explained to the Terrestrials his own mode of life, and received from them in turn full information concerning Earthly life, activity, and science. Long they talked, and it was almost time to slow down for the journey's end when the Callistonian brought the conversation back to their immediate concerns.

"My lieutenant and I were upon a mission of some importance, but it is more important to take you to Callisto, for there may be many things in which you can help us. Not in rays—we know all the vibrations you have mentioned and several others. The enemy, however, is supreme in that field, and until our scientists have succeeded in developing ray-screens, such as are used by the hexans, it would be suicidal to use rays at all. Such screens necessitate the projection of pure, yet dirigible, forces—you do not have them upon your planet?"

"No, and so far as I know such screens are also unknown upon Mars and Venus, with whose inhabitants we are friendly."

"The inhabitants of all the planets should be friendly; the solar system should be linked together in intercourse for common advancement. But that is not to be. The hexans will eventually triumph here, and a Jovian system peopled by hexans will have no intercourse with any human civilization save that of internecine war. We, of Callisto, have only one hope—or is it really a hope? In the South Polar country of Jupiter, there dwells a race of beings implacably hostile to the hexans. They seem to invade the country of the hexans frequently, even though they are apparently repulsed each time. Our emissaries to the South Polar country, however, have never returned—those beings, whatever they are, if not actively inimical, certainly are not friendly toward us."

"You know nothing of their nature?"

"Nothing, since our electrical instruments are not sufficiently sensitive to give us more than a general idea of what is transpiring there, and vision is practically useless in that eternal fog. We know, however, that they are far advanced in science, and we are thankful indeed that none of their frightful flying fortresses have been launched against us. They apparently are not interested in the satellites, and it is no doubt due to their unintentional assistance that we have survived as long as we have."

In the cavern at last, the three men boarded the Callistonian space-plane and were shot up the crater's shaft. The voyage to Callisto was uneventful, even uninteresting save at its termination. The *Bzarvk*, coated every inch as it was with a dull, dead black, completely absorptive outer coating, entered the thin layer of Callisto's atmosphere in darkest night, with all rockets dead, with not a light showing, and with no apparatus of any kind functioning. Utterly invisible and undetectable, she dove downward, and not until she was well below the crater's rim did the forward rockets burst into furious life. Then the Terrestrials understood another reason for the immense depth of those shafts other than that of protection from the detectors of the enemy—all that distance was necessary to overcome the velocity of their free fall without employing a negative acceleration greater than the frail Callistonian bodies could endure. From the cavern at the foot of the shaft, a regulation tunnel extended to the Callistonian city of Zbardk.

Portal and city were very like Wruszk, upon distant Europa, and soon the terrestrial captain and pilot were in conference with the Council of Callisto.

MONTHS of Earthly time dragged slowly past, months during which King and Breckenridge studied intensively the offensive and defensive systems of Callisto without finding any particular in which they could improve them to any considerable degree. Captain Czuv and his warplane still survived, and it was while the Callistonian commander was visiting his terrestrial guests, that King voiced the discontent that had long affected both men.

"We're both tired of doing nothing, Czuv. We have been of little real benefit, and we have decided that your ideas of us are all wrong. We are convinced that our personal horsepower can be of vastly more use to you than our brain-power, which doesn't amount to much. Your whole present policy is one of hiding and sniping. I think that I know why, but I want to be sure. Your vessels carry lots of fuel—why can the hexans outrun you?" Thus did King put his problem.

"They can stand enormously higher accelerations than we can. The very strongest of us loses consciousness at an acceleration of twenty-five meters per second per second, no matter how he is braced, and that is only a little greater than the normal gravity of our enemies upon Jupiter. Their vessels at highest power develop an acceleration of thirty-five meters, and the hexans themselves can stand much more than even that high figure," replied Czuv.

"I thought so. Assume that you traveled at forty-five. Would it disable you permanently, or would you recover as soon as it was lowered?"

"We would recover promptly, unless the exposure had been unduly prolonged. Why?"

"Because," said King, "I can stand an acceleration of fifty-four meters for two hours, and Breckenridge here tests fifty two meters. I can navigate anything, and Breckenridge can observe as well as any of your own men. Build a plane to accelerate at forty-five meters and we will blow those hexans out of the ether. You will have to revive and do the shooting, however—your gunnery is entirely beyond us."

"That is an idea of promise, and one that had not occurred to any of us," Czuv replied and work was begun at once upon the new flyer.

When the super-plane was ready for its maiden voyage, its crew of three studied it as it lay in the catapult at the portal. Dead black as were all the warplanes, its body was twice as large as that of the ordinary vessel, its wings were even more stubby, and its accommodations had been cut to a minimum to make room for the enormous stores of fuel necessary to drive the greatly increased battery of rocket motors and for the extra supply of torpedoes carried. Waving to the group of soldiers and citizens gathered to witness the take-off of the new dreadnought of space, the three men entered the cramped operating compartment, strapped themselves into their seats, and were shot away. As usual the driving rockets were cut off well below the rim of the shaft, and the vessel rose in a long and graceful curve, invisible in the night. Such was its initial velocity and so slight was the force of gravity of the satellite that they were many hundreds of miles from the exit before they began to descend, and Breckenridge studied his screens narrowly for signs of hexan activity.

"Do you want to try one of your long-range shots when we find one of them?" the pilot asked Czuv.

"No, it would be useless. Between deflection by air-currents and the dodging of the enemy vessels, our effective range is shortened to a few kilometers, and their beams are deadly at that distance. No, our best course is to follow the original plan—to lure them out into space at uniform acceleration, where we can destroy them easily."

"Right," and Breckenridge turned to King, who was frowning at his controls. "How does she work on a dead stick, Chief?"

"Maneuverability about minus ten at this speed and in this air. She'd have to have at least fifteen hundred kilometers an hour to be responsive out here. See anything yet?"

"Not yet ... wait a minute! Yes, there's one now—P-12 on area five. Give us all the X10 and W27 you can, without using power—we want to edge over close enough so that she can't help but see us when we start the rockets."

"Be sure and stay well out of range. I'm giving her all she'll take, but she won't take much. With these wings she has the gliding angle of a kitchen sink."

"All x—I'm watching the range, close. Wish we had instruments like these on the IPV's. We'll have to install some when we get back. All x! Give her the gun—level and dead ahead!"

Half the battery of rockets burst into their stuttering, explosive roar of power and the vessel darted away in headlong flight.

"He sees us and is after us—turn her straight up!"

A searing, coruscating finger of flame leaped toward them, but their calculations had been sound—the hexan was harmless at that extreme range. King, under the pilot's direction, kept the plane at a safe distance from the sphere while the satellite grew smaller and smaller behind them and Czuv lapsed quietly into unconsciousness.

"He's been out for quite a while. Far enough?" asked King.

"All x now, I guess—don't believe they can see the flash from here. Cut!"

The rockets died abruptly and a blast from the side ports threw the plane out of the beam—and once out of it, beyond range of the electromagnetic detectors as they were their coating of absolute black rendered the craft safe from observation. One dirigible rocket remained in action, its exhaust hidden from the enemy by the body of the vessel, and Captain Czuv soon recovered his senses.

"Wonderful, gentlemen!" he exclaimed, as he manipulated the delicate controls of his gunnery panel. "This is the first time in history that a Callistonian vessel has escaped from a hexan by speed alone."

An instantaneously extinguished flare of incandescence marked the passing of the hexan sphere into nothingness, and the cruiser shot back toward Callisto in search of more prey. It was all too plentiful, and twenty times the drama was reenacted before approaching day made it necessary for Czuv to take the controls and dive the vessel into the westermost landing-shaft of Zbardk. A rousing and enthusiastic welcome awaited them, and joy spread rapidly when their success became known.

"Now we know what to do, and we had better do it immediately, before they get our system figured out and increase their own power." King reported to the Council. "You might send a couple of ships to Europa and bring back as many of the Tellurian officers as want to come and can be spared from the work there. They all test above forty-five meters, and they can learn this stuff in short order. While they're coming, your engineers can be building more ships like this one."

The new vessel did not make another voyage until nine sister ships were ready and manned, each with two Terrestrial officers and one Callistonian gunner. All ten took to the ether at once, and the hexan fleet melted away like frost-crystals before a summer sun. A few weeks of carnage and destruction and not a hexan was within range of the detectors of Callisto—they were gone!

"This is the first time in years that Callisto's air has been free of the hexans," Czuv said, thoughtfully. "With your help we have reduced their strength to a fraction of what it was, but they have not given up. They will return, with a higher acceleration than even you Terrestrials, powerful as you are, can stand."

"Certainly they will, but you will be no worse off than you were before—you can return to your own highly effective tactics."

"We are infinitely better off for your help. You have given us a new lease on life...."

He broke off as a flaring light sprang into being upon the portal board and the observer of Exit One made his report—there was a hexan vessel in the air, location 425 over VJ-42.

"There's one left! Let us get him! No, he's ours!" Confused shouts arose from the bull-pen; but the original superplane was at the top of the call-board and accordingly King, Breckenridge, and Czuv embarked upon an expedition more hazardous far than they had supposed—an expedition whose every feature was relayed to those in the portal by the automatic lookouts upon the rims and which was ended before a single supporting Callistonian plane could be launched.

For the enemy vessel was not the last of the low-powered hexan vessels, as everyone had supposed—it was the first of the high-powered craft, arriving long before its appearance was expected. Before its terrific acceleration and savage onslaught, the superplane might as well have been stationary and unarmed. After his long dive downward, King could not even leave the atmosphere—the hexan was upon them within a few seconds, even though the stupendous battery of rockets, full driven, had roared almost instantly into desperate action. Bomb after bomb Breckenridge hurled, with full radio control, fighting with every resource at his command, but in vain. The frightful torpedoes were annihilated in mid-flight; and nose, tail-assembly, and wings were sheared neatly from the warplane by a sizzling plane of force. Side rockets and torpedo tubes were likewise sliced away and the helpless body of the Callistonian cruiser, falling like a plummet, was caught and held by a tractor ray. Captor and captive settled toward the ground.

"This is a signal honor," observed Captain Czuv when he had revived. "It has been many, many cycles since they have taken Callistonians captive. They kill us at every opportunity. Is it your custom to destroy yourselves in a situation such as this?"

"It is not. While we live there is hope."

"Not ours. Unless they have made enormous strides in psychological mechanisms, they cannot tear from our minds any secrets we really wish to keep. That is useless," he went on, as King lifted a hand-weapon. "You will have no opportunity whatever to use it," and he was right.

A searing beam of energy drove them out of the vessel, then electromagnetic waves burned every metallic object out of their possession.

Burning rays herded them into the hexan sphere and into a small room, whose door clanged shut behind them.

"Ah, two are humans of a strange breed!" a snarling voice barked from the wall, in the Callistonian language. "Our deductions were accurate, as usual—it is to the humans of Planet Three, whose bodies are a trifle less puny than those of the humanity of the satellites, that we owe our recent reverses. However, those reverses were merely temporary—humanity, no matter what its breed, shall very shortly disappear from the satellites. Now, you scum of the Solar System, you shall be permitted to witness an entrancing spectacle on the way to our headquarters, where all your knowledge is to be taken from you before you die, lingeringly and horribly. There is a strange space-vessel nearing us probably searching for the one we took and which you dogs of Callisto must have been fortunate enough to take from us before we could study and kill its human cargo. Watch its destruction and cringe—and know, in your suffering, that the more you suffer, the greater shall be our enjoyment."

"I believe that," King acknowledged. As all three prisoners stared at the wall-screen, upon which was pictured a huge football of scarred grey steel, Czuv was amazed to see the faces of Breckenridge and King light up with fierce smiles of pleasure and anticipation.

"You dissemble well," remarked the Callistonian. "That will rob them of much pleasure."

"They'll get robbed of more than that," King returned. "This is too good to keep, and since they cannot understand English, I'll tell you something. I told you about Stevens. He apparently wasn't killed, as we thought. He must have escaped, and there is the result. That ship there is far from innocent—her being so far out of range of any of our power-plants proves that. That vessel is the *Sirius*—the research laboratory of the IPC—the Inter-Planetary Corporation! It carries the greatest scientific minds of three of the inner planets, and it is loaded with pure poison or it wouldn't be here. Oh, you hexans, what you have got coming to you!"

Concluding a Thrilling New Serial of Interplanetary Life and Travel by Edward E. Smith, Ph.D.

Author of "Skylark of Space," and "Skylark Three"

PART III

THE question of rays – their expanding power for good and evil – is receiving increasing attention from scientists. The x-ray has been found to be very beneficial, given in certain quantities, but extremely inimical to health, and even fatal, if too much exposure is given. The powers of the cosmic rays have not been fully discovered as yet. And there is no reason to doubt the theory that there may be found still more destructive and powerful rays. Even wars are becoming a more dangerous plaything for nations of our world – to say nothing of other possible enemies from other parts of our universe. Stevens and Nadia Newton meet with thrilling experiences galore in this concluding instalment.

What Went Before:

THE Interplanetary Vessel *Arcturus* sets out for Mars, with Breckenridge as chief pilot, carrying on board, besides its regular crew and some passengers, the famous Dr. Stevens, designer of space ships and computer. He checks computations made by astronomers stationed in floating observatories, and after he has located any trouble and suggests a plan for minimizing the hazards of the trip from the earth to Mars he reports his findings and suggestions to Mr. Newton, chief of the Interplanetary Corporation.

Stevens then takes Nadia, Mr. Newton's beautiful young daughter, on a specially conducted sight-seeing tour of the *Arcturus* and thoroughly explains to her all of the works of the vessel. Nadia has herself had a good science education. While they are down at the bottom of the ship—nearing the end of their tour—Stevens feels a barely perceptible movement of the vessel from its course. When he turns on the visiplate, he is horrified to find that a mysterious ray of unparalleled power has neatly sliced the *Arcturus* in several places.

Nadia and Stevens are completely separated from the rest of the crew and passengers of the ship, so they get into a lifeboat, which is equipped for a limited amount of space travel. Despite the strict and apparently effective vigilance of the enemy destroyer, Stevens and Nadia make their getaway in the lifeboat, which they aptly call "Forlorn Hope," and finally make a safe landing on Ganymede, where Stevens almost completes a power-plant and a radio transmitter, to enable him to communicate with the earth or with the IPV *Sirius*, which is used by Westfall and Brandon (two of the world's best scientists) as a floating laboratory.

They start for Cantrell's Comet, where Stevens acquires the necessary material for his giant transmitting tube, heads back to Ganymede, when their ship is cut, top and bottom, by a strong ray-beam. Stevens and Nadia soon find that the other ship is manned by friendly beings from Saturn. Together they plan against their common foes—the Hexans—who are enemies of the universe. After helping the Saturnians to repair their power plant, they start back to Ganymede, aided by their new friends from the frigid civilization. Finally, however, Stevens succeeds in connecting, by radio, with the *Sirius* and his scientist friends on board it, who rush to the aid of the two castaways. It is while the castaways are captives of the Hexans that help looms near.

CHAPTER IX
The Sirius Takes a Hand

THE *Sirius* loafed along through the ether at normal acceleration just outside the orbit of Mars and a million miles north of the ecliptic plane. In the control room, which had been transformed into a bewilderingly complete laboratory, Norman Brandon strode up and down, waving his arms, his unruly black hair on end, addressing savagely his friend and fellow-scientist, who sat unmoved and at ease.

"For cat's sake, Quince, let's get busy! They're outside somewhere, since the police have scoured every cubic kilometer within range of the power plants without finding a trace of them. We've got the power question licked right now—with these fields we can draw sixty thousand kilofranks from cosmic radiation, which is lots more than we'll ever need. We haven't drawn a frank from a plant in a month, and we've had to cut our field strength down to a whisper to keep from burning out our accumulators. We can hunt as far as Neptune easy—we can go to Alpha Centauri if we want to. This thing of piffling and monkeying around here's pulling my cork, and for the ten thousand four hundred and sixty seventh time I say *let's prowl* and *prowl now*! In fact, I'm getting so sick of sticking around doing nothing that I'm going out anyway, if I have to go alone in a lifeboat!"

The flying fortresses were finally wrenched from the ground and hurled upward.

Impetuous and violent as Brandon had always been, never before had he gone to such lengths as to suggest a disruption of the partnership; and Westfall, knowing that Brandon, in his most violent moments, never threatened idly, thought long before he replied.

"You will not go alone, of course. If you insist upon going without further preparation I will go too, no matter how foolish I think such a course to be. We have power, it is true, but in all other respects we are in no condition to meet an opponent having command of such resources as must certainly be possessed by those who attacked the *Arcturus*. Our detectors are inefficient, our system of vision is crude, to say the least, and many other things are still in the experimental stage. We have not the slightest idea whom or what we may encounter. It is all too probable that we would simply be throwing away uselessly the lives of more good men. It is also foolish from a general viewpoint, for as you already know, we and our assistants happen to be in better position to study these things than is any one else at the present time. However, I will compromise with you. We can learn much in a month if you will really try, instead of wasting time in fuming around the ship and indulging in these idiotic tantrums. If you will buckle down and really study the problems confronting us for thirty days, we will set out at the end of that time, ready or not."

"All x. I hate to do it, but we've been together too long to bust it up now," and Brandon turned toward his bench. Scarcely had he reached it when a series of dots and dashes roared from an amplifier. Both men leaped for the receiver which had so unexpectedly burst into sound, reaching it just as it relapsed into silence, and from the tape of the recorder they read the brief message.

"...h four seven ganymede point oh four seve...."

"That's Steve!" yelled Brandon. "Nobody else could build an ultra-sender! Direction?"

"No need of calculating distance or direction. Ganymede is the third major satellite of Jupiter."

"Sure. Of course, Quince—never thought of that. Dope enough—point oh four seven."

As Stevens had told Nadia, the message was completely informing to those for whom it was intended, and soon Brandon's answer was flying toward the distant satellite. He then started to call the officers of the Interplanetary Corporation, but was restrained by his conservative friend.

"It would be better to wait a while, Norman. In a few hours we will know what to tell them."

At high acceleration the *Sirius* drove toward the Jupiter-Earth-North plane, and Brandon calculated from his own bearings and from the current issue of the "Ephemeris" the time at which Stevens' reply should be received. Two minutes before that time he was pacing up and down in front of the ultra-receiver, and fifteen seconds after it he snapped:

"Come on, Perce, get busy! Shake a leg!"

"Oh, come, Norman; give him a few minutes' leeway, at least," said Westfall, with amused tolerance. "Even if your calculations are that accurate—which of course they are," he added hastily at a stormy glance from hot black eyes, "since we received that message direct, instead of through one of our relay stations, Stevens probably has been throwing it around for hours or perhaps days, looking for us, and the shock of hearing from us at last might well have put him out of control for a minute or two."

The carrier wave hissed into the receiver, forestalling Brandon's fiery reply, followed closely by the code signals they had been expecting. As soon as the story had been told, and while Brandon was absorbed in the scientific addenda of Stevens, Westfall thoughtfully called up Newton, Nadia's father.

"Nadia is alive, free, safe, well, and happy," he shot out without preliminary or greeting, as soon as the now lined features of the director showed upon the communicator screen, and the careworn countenance smoothed magically into the keen face of the fighting Newton of old, as Westfall recounted rapidly the tale of the castaways.

"They apparently have not suffered in any way," he concluded. "All that Stevens wants is some cigarettes, and your daughter's needs, while somewhat more numerous than his, seem to be only clothes, powder, perfume, and candy. Therefore we need not worry about them. The fate of the others is still unknown, but there seems to be a slight possibility that some of them may yet be rescued. You may release as much or as little of this story as may seem desirable. Stevens is still sending data of a highly technical nature. We shall arrive there at 21:32 next Tuesday."

IN due time the message from Ganymede ended and Brandon, with many pages of his notebook crammed with figures and equations, snapped off the power of the receiver and turned to his bench. Gone was the storming, impetuous rebel; his body was ruled solely by the precise and insatiable brain of the research scientist.

"He's great, that kid Perce! When I see him, I'm going to kiss him on both cheeks. He's got enough dope on them to hang them higher than Franklin's kite, and we'll nail those jaspers to the cross or I'm a polyp! He's crazier than a loon in most of his hunches, but he's filled four of our biggest gaps. There is such a thing, as a ray-screen, you kill-joy, and there are also lifting or tractor rays—two things I've been trying to dope out and that you've been giving me the Bronx cheer on. The Titanians have had a tractor ray for ages—he sent me complete dope on it—and the Jovians have got them both. We'll have them in three days, and it ought to be fairly simple to dope out the opposite of a tractor, too—a pusher or presser beam. Say, round up the gang, will you, while I'm licking some of this stuff into shape for you to tear apart? Where are Venus and Mars? Um ... m ... m. Tell Alcantro and Fedanzo to come over here pronto—give 'em a special if necessary. We'll pick up Dol Kenor and Pyraz Amonar on the way—no, get them to Tellus, too. Then we'll get action quicker. Those four are all I want—get anybody else you want to come along."

His hands playing over the keys of an enormous calculating machine, Brandon was instantly immersed in a profound mathematico-physical problem; deaf and blind to everything about him. Westfall, knowing well that far-reaching results would follow Brandon's characteristic attack, sat down at the controls of the communicator. He first called Mars, the home planet of Alcantro and Fedanzo, the foremost force-field experts of three planets; and was assured in no uncertain terms that those rulers of rays were ready and anxious to follow wherever Brandon and Westfall might lead. Thence to Venus, where Dol Kenor, the electrical wizard, and Pyraz Amonar, the master of mechanism, also readily agreed to accompany the expedition. He then called the General-in-Chief of the Interplanetary Police, requesting a detail of two hundred picked men for the hazardous venture. These most important calls out of the way, he was busy for over an hour giving long-distance instructions so that everything would be in readiness for the servicing of the immense space-cruiser the following Tuesday night.

Having guarded against everything his cautious and far-seeing mind could envisage, he went over to Brandon's desk and sat down, smoking contemplatively until the idea had been roughed out in mathematical terms.

"Here's the rough draft of the ray screen, Quince. We generate a blanket frequency, impressed upon the ultra carrier wave. That's old stuff, of course. Here's the novelty, in equation 59. With two fields of force, set up from data 27 to 43, it will be possible actually to project a pure force of such a nature that it will react to de-heterodyne the blanketing frequency at any predetermined distance. That, of course, sets up a barrier against any frequency of the blanketed band. Incidentally, an extension of the same idea will enable us to see anywhere we want to look—calculate a retransmitting field."

"One thing at a time, please. That screen may be possible, but those fields will never generate it. Look at datum 31, in which your assumptions are unsound. In order to make any solution at all possible you have assumed cosine squared theta negligible. Mathematically, it is of course vanishingly small compared to the first power of the cosine, but fields of that type must be *exact*, and your neglect of the square is indefensible. Since you cannot integrate with the squared term in place, your whole solution fails."

"Not necessarily. We'll go back to 29, and put in sine squared theta minus one equal to z sub four. That gives us a coversed sine in 30, and then we integrate...."

Thus the argument raged, and all the assistants whose work was not too pressing gathered around unobtrusively, for it was from just such fierce discussions as this that the ultra-radio and other epoch-making discoveries had come into being. Yard after yard of calculator paper was filled with equations and computations. Weirdly shaped curves were drawn, with arguments at every point—arguments hot and violent from Brandon, from Westfall cold and precise, backed by lightning calculations and with facts and diagrams culled from the many abstruse works of reference, which by this time literally covered the bench and overflowed upon the floor.

It was in this work that the strikingly different temperaments and abilities of the two scientists were revealed. Brandon never stood still, but walked around jerkily, chewing savagely the stem of an ancient and reeking pipe, gesticulating vigorously, the while his keen and agile mind was finding a way over, around, or through the apparently insuperable obstacles which beset their path; by means of mathematical and physical improvisations, which no one not inspired by sheer genius could have evolved. Westfall, seated quietly at the calculator, mercilessly shredded Brandon's theories to ribbons, pointing out their many flaws with his cold, incisive reasoning and with rapid calculations of the many factors involved. Then Brandon would find a remedy for each weakness in turn and, when Westfall could no longer find a single flaw in the structure, they would toss the completed problem upon a table and attack the next one with unabated zeal. Brandon, in his light remark that the two made one real scientist, had far understated the case—those two brains, each so powerful and each so perfectly complementing the other, comprised the master-scientist who was to revolutionize science completely in a few short years.

To such good purpose did they labor that the calculations were practically finished by the time they reached the earth. There the ship was serviced with a celerity that spoke volumes for the importance of her mission—even the *Aldebaran*, the dazzlingly gold-plated queen of the fleet, waited unattended and disregarded on minus time while the entire force of the Interplanetary Corporation concentrated upon the battle-scarred old hulk of the *Sirius*. Brandon was surprised when he saw the two companies of police, but characteristically accepted without question the wisdom of any decision of his friend, and cordially greeted Inspector-General Crowninshield, only a year or so older than himself, but already in charge of a Division.

"Keen-looking bunch, Crown. Lot of different outfits—volunteers for special duty from the whole Tellurian force?"

"Yes. Everybody wanted to go, and there threatened to be trouble over the selection, so we picked the highest ratings from the whole Service. If there ever was such a thing as a picked force, we shall have it with us."

"What d'you mean, 'us'? You aren't going, are you?"

"Try to keep me from it! The names of all five of us I-G's were put in a hat, and I was lucky."

"Well, you may come in handy, at that," Brandon conceded. "And here's the big boss himself. Hi, Chief!"

"Ho, Brandon! Ho, Westfall!" Newton, Chairman of the Board of Directors of the IPC, shook hands with the two scientists. "Your Martians and Venerians are in Lounge Fifteen. I suppose that you have a lot of things to thrash out, so you may as well start now. Everything is being attended to—I'll take charge now."

"You going along, too?" asked Brandon.

"Going along, *too*? I'm *running* this cruise!" Newton declared. "I may take advice from you on some things and from Crowninshield on others, but I am in charge!"

"All x—it's a relief, at that," and Brandon and Westfall went to join their fellow-scientists in the designated room of the space-cruiser.

WHAT a contrast was there as the representatives of three worlds met! All six men were of the same original stock or of a similar evolution—science has not, even yet, decided the question definitely. Their minds were very much alike, but their respective environments had so variantly developed their bodily structures that to outward seeming they had but little in common.

Through countless thousands of generations the Martians had become acclimated to a planet having little air, less water, and characterized by abrupt transitions from searing heat to bitter cold: from blinding light to almost impenetrable darkness. Eight feet tall and correspondingly massive, they could barely stand against the gravitational force of the Earth, almost three times as great as that of Mars, but the two Martian scientists struggled to their feet as the Terrestrials entered.

"As you were, fellows—lie down again and take it easy." Brandon suggested in the common Interplanetarian tongue. "We'll be away from here very soon, then we can ease off."

"We greet our friends standing as long as we can stand," and, towering a full two feet above Brandon's own six-feet-two, Alcantro and Fedanzo in turn engulfed his comparatively tiny hand in a thick-shelled paw and lifted briefly the inner lids of quadruply-shielded eyes. For the Martian skin is not like ours. It is of incredible thickness; dry, pliable, rubbery, and utterly without sensation: heavily lined with fat and filled throughout its volume with tiny air-cells which make it an almost perfect non-conductor of heat and which prevent absolutely the evaporation of the precious moisture of the body. For the same reasons their huge and cat-like eyes are never exposed, but look through sealed, clear windows of membrane, over which may be drawn at will one or all of four pairs of lids—lids transparent, insensible, non-freezable, air-spaced insulators. Even the air they exhale carries from their bodies a minimum of the all-important heat and moisture, for the passages of their nostrils do not lead directly to the lungs, as do ours. They are merely the intakes for a tortuous system of tubes comprising a veritable heat-exchanger, so that the air finally expelled is in almost perfect equilibrium with the incoming supply in temperature and in moisture content. A grayish tan in color, naked and hairless—though now, out of deference to Terrestrial conventions, wearing light robes of silk—indifferent alike to any extreme of heat or cold, light or darkness: such were the two forbidding beings who arose to greet their Terrestrial friends, then again reclined.

"I suppose that you have been given to drink?" Westfall made sure that they had been tendered the highest hospitality of Mars.

"We have drunk full deeply, thanks; and it was not really necessary, for we drank scarcely three weeks ago."

Brandon and Westfall turned then and greeted the two Venerians, as different from the Martians as they were from the Terrestrials. Of earthly stature, form, and strength, yet each was encased in a space-suit stretched like a drum-head, and would live therein or in the special Venerian rooms of the vessel as long as the journey should endure. For the atmosphere of Venus is more than twice as dense as ours, is practically saturated with water-vapor, carries an extremely high concentration of carbon dioxide, and in their suits and rooms is held at a temperature of one hundred and ten degrees Fahrenheit. The lenses of their helmets were of heavy, yellowish-red composition, protecting their dead-white skins and red eyes from all actinic rays—for the Venerian lives upon the bottom of an everlasting sea of fog and his thin epidermis, utterly without pigmentation, burns and blisters as frightfully at the least exposure to actinic light as does ours at the touch of a red-hot iron.

Out in space at last, cruising idly with the acceleration set at a point bearable for the Martians, Westfall called the meeting to order and outlined the situation facing them. Brandon then handed around folios of papers, upon which the Venerians turned the invisible infra-red beams of the illuminators upon their helmets, thus flooding them with the "light" to which their retinas were most responsive.

"Here's the data," Brandon began. "As you see from Sheet 1, we can already draw any amount of power we shall need from cosmic radiation alone...."

"Perpetual motion—ridiculous!" snapped from the sending disk upon the helmet of the master of mechanism.

"Not at all, Amonar," put in his fellow Venerian, "any more than a turbo-generator at the foot of a waterfall is perpetual motion. Those radiations originate we know not where, probably as a result of intra-atomic reactions. The fields of force of our hosts merely intercept these radiations, as a water-driven turbine intercepts the water. We merely use a portion of their energy before permitting them to go on, to we know not what end. Truly you have made a notable achievement in science, Tellurian friends, and we congratulate you upon its accomplishment. Please proceed."

"Upon the following sheets are described the forces employed by the Jovians, as we shall call them until we find out who or what they really are. We will discuss these forces later. For each force we have already calculated a screen, and we have also calculated various other forces of our own, with which we hope to arm ourselves before we reach Ganymede. The problems facing us are complex, since there are some nine thousand forcebands of the order in which we are working, each differing from all the others as much as torque differs from tension, or as much as red differs from green. Therefore we have appealed to you for help, knowing that we could do but little alone.

Alcantro and Fedanzo will supervise the construction of the generators of the various fields from these calculations. Dol Kenor will correlate power and electricity to and with the fields. Westfall and I will help work out the theoretical difficulties as they arise. Pyraz Amonar, who can devise and build a machine to perform any conceivable mechanical task, will help us all in the many mechanical difficulties we shall certainly encounter. Discussion of any point is now in order."

STEP by step and equation after equation the calculations and plans were gone over, until every detail was clear in each mind. Then the men bent to their tasks; behind them not only the extraordinarily complete facilities of that gigantic workshop which was the *Sirius*; but also the full power of the detachment of police—the very cream of the young manhood of the planet. Week after toilsome week the unremitting labor went on, and little by little the massive cruiser of the void became endowed with an offensive and defensive armament incredible. An armament conceived in the fertile and daring brain of a sheer genius, guided only by the knowledge that such things were already in existence somewhere; reduced to working theory by a precise, mathematical logician; translated into fields of force by the greatest known experts; powered by the indefatigable efforts of an electrical wizard; made possible by the artful mechanical devices of the greatest inventor that three worlds had ever known! Thus it was that they approached Ganymede, ready, with blanketing screens full out, save for one narrow working band, and with a keen-eyed observer at every plate. When even the hyper-critical Westfall was convinced that their preparations were as complete as they could be made with the limited information at hand, Brandon directed a beam upon the satellite and tapped off a brief message:

"stevens ganymede will arrive in about ten hours direct carrier beam toward sun we can detect it and will follow it to wherever you are sirius."

"ipv sirius," came the reply, "everything here, all x glad to see you thanks newton and stevens."

Brandon, at the controls, scanning his screens narrowly, dropped the vessel down to within a mile or two of the point of origin of Stevens' carrier beam without incident; then spoke to Westfall, at his side, with a grin.

"Nice layout the kid's got down there, Quince. It's too bad—don't look like we're going to get any action for our money a-tall. 'Sa shame, too—what's the use of wasting it, now that we've got it all made?"

"We are not done yet," cautioned Westfall, and even as he spoke an alarm bell burst into strident clamor—one of their far-flung detector screens was telling the world that it had encountered a dangerous frequency. The new ultra-lights flared instantly along the line automatically laid down by the detector, and upon the closely ruled micrometer screen of Brandon's desk there glowed in natural color the image of a globular space-ship, approaching them with terrific speed.

"Men all stationed, of course, Crown?"

"Stationed and ready." Crowninshield, phones at his ears and microphone at his lips, was staring intently into his own plate.

"Kinda think I'll do most of it from here, but you can't always tell. If they get inside my guard you all know what to do."

"All x."

Expecting another such hollow victory as the other Hexan vessel had won over the defenseless *Arcturus*, the small stranger flashed nearer and nearer that huge and featureless football of armor steel. Within range, she launched her flaming plane of energy, but this time that Jovian sheet of force did not encounter unprotected and non-resisting steel. Upon the outer ray-screen, flaming white into incandescent defense, the furious bolt spent itself, and in the instant of the launching of the searing blade of flame, Brandon had gone into action. Switch after switch drove home, and one after another those frightful fields of force, those products of the mightiest minds of three planets, were hurled out against the tiny Jovian sphere. Driven as they were by the millions upon millions of horsepower stored in the accumulators of the *Sirius* they formed a coruscating spherical shell of intolerable energy all around the enemy vessel, but even their prodigious force was held at bay by the powerful defensive screens of the smaller space-ship. But attack the Jovian could not, every resource at her command being necessary to fend off the terrific counter-attack of her intended prey, and she turned in flight. Small and agile as she was, the enormous mass of the *Sirius* precluded any possibility of maneuvering with the Jovian, but Brandon had no intention of maneuvering. Rapid as the motions of the stranger were and frantic as was her dodging, the terrific forces of the tractor beams of the Interplanetary Vessel held her in an unbreakable grip, and although she dragged the massive *Sirius* hither and thither, she could not escape.

"Hm ... m ... m," mused Brandon. "We seem to be getting nowhere fast. How much power we using, Mac, and how much have we got coming in?"

"Output eighty-five thousand kilofranks," replied MacDonald, the first assistant. "Intake forty-nine thousand."

"Not so good—can't hold out forever at that rate. Shove out the receptor screens to the limit and drive 'em. They figure a top of sixty thousand, but we ought to pick up a little extra from that blaze out there. Drive 'em full out or up to sixty-five, whichever comes first. Can't seem to crush his screens, so

I guess we'll have to try something else," and a thoughtful expression came over his face as he slowly extended his hand toward another switch, with a questioning glance at Westfall.

"Better not do that yet, Norman. Use that only as a last resort, after everything else has failed."

"Yeah—I'm scared to death of trying it, and it isn't necessary yet. He must have an open slit somewhere to work through, just as we have. I'll feel around for it a while."

"Is there any way of hetrodyning the new visiray upon the exploring frequency?"

"Hm ... m.... Never thought of that—it would be nice, too.... I think we can do it, too. Watch 'em, Quince, and holler if they start anything."

He abandoned his desk and established the necessary connections between the visiray apparatus and the controls of his board. There was a fierce violet-white glare from the plate as he closed the switch, and he leaped back with his hands over his eyes, temporarily blinded.

"Wow, that's hot stuff!" he exclaimed. "It works, all x, to the queen's taste," as he donned his heavy ray-goggles and resumed his place.

After making certain that the visiray was precisely synchronized and phased with the searching frequency, he built up the power of that beam until it was using twenty thousand kilofranks. Then, by delicately manipulating the variable condensers and inductances of his sensitive shunting relay circuits, he slowly shifted that frightful rod of energy from frequency to frequency, staring into the brilliant blankness of his micrometer screen as he did so. After a few minutes of search the screen darkened somewhat, revealing the image of the Jovian globe. Brandon instantly shifted into that one channel the entire power of his attack; steadying the controls to bring the sphere of the Jovians into the sharpest possible focus, knowing that he had found the open slit and that through it there was pouring upon the enemy the full power of his terrible weapon.

In the fraction of a second before the Jovians could detect the attack and close the slit, he saw a portion of the wall of their vessel flare into white heat and literally explode outward in puffs and gouts of flaming, molten metal and of incandescent gases. But the thrust, savage as it was, had not been fatal and the enemy countered instantly. Now that the crushing force of the full-coverage attack was lessened for a moment, through another slit there poured a beam of energy equal to the Terrestrials' own—a beam of such intense power that the outer screen of the *Sirius* flared from red through the spectrum, to and beyond the violet, and went black in less than a second, and the inner screen had almost gone down before Brandon's lightning hands could restore the complete coverage that so effectively blanketed the forces of the enemy.

"Well, we're back to the *status quo,*" announced Brandon, calmly. "It's a good gag they didn't have time to locate our working slit—if they had pushed that stuff through our open channel, we'd have gotten frizzled up some around the edges. As it was, we got the edge on that exchange—take it from your Uncle Dudley, Quince, that bird knows that he's been nudged!"

AGAIN he searched the entire band for an opening, but could find none. The enemy had apparently retired into a tightly closed shell of energy. The small vessel no longer struggled, nor even moved, but was merely resisting passively.

"Not an open channel, not even one for him to work through—he can't wiggle. Well, that won't get him anything. We're so much bigger than he is, that we can outlast him and will get him some time, since he's bound to run out of power before we do. I don't believe he can receive anything, sealed up as he is, and he can't have accumulators enough more efficient than ours to make up the difference, can he, Quince?"

"It is quite possible. For instance, although we have never heard of any progress being made along such lines, it has been pointed out repeatedly that synthesis of a radio-active element of very high atomic weight would theoretically yield an almost perfect accumulator—one many thousands of times as efficient as ours in mass-to-energy ratio. Then, too, you realize, of course, that there is a bare possibility that intra-atomic energy may not be absolutely impossible."

"Nix on that, Quince. I'll stand for a lot, but not for that last idea! It's hard to say that anything's impossible, of course, except things made so by definition or by being contrary to observational facts, but the best work shows that intra-atomic energy is just about as impossible as anything can well be. It has been shown pretty conclusively that all ordinary matter is already in its most stable state, so that work must be done upon any ordinary atom to decompose it. Besides, if he had either radioactive accumulators or intra-atomic energy, he would have cut us up long ago. Nope, the answer is that he's probably yelled for help and is trying to hold out until it gets here," was Brandon's rejoinder.

"What can we do about it?" asked Quince.

"Don't know yet. I do know, though, that we aren't half as ready for trouble as I thought we were. There's a dozen things I want to do that I can't because we haven't got the stuff. Don't say 'I told you so,' either—I know you did! You're the champion ground-and-lofty thinker of the century. Alcantro!"

"Here!"

"Round up the gang, will you, and figure me out a screen and a set of meters that will indicate an open band? We lose too much time feeling around anyhow, and we're too apt to take one on the chin while we're doing it. Also, you ought to make it so it'll shoot a jolt into the opening, while you're at it," said Brandon.

"We shall begin at once," and the massive Martian as he replied, stepped over to the calculating machine.

"Well, Quince, we can't do much to him this way—he's crawled into a hole and pulled the hole in after him. Gosh, I wish we had more stuff!"

"After all, we have everything whose necessity and practicability could have been foreseen in the light of our information. We can, of course, go further."

"You chirped it! But we can't let things ride this way or we'll get our hair singed. We'll have to decorate him with the grand slam, I guess."

"Yes, it seems as though the time for emergency measures has arrived."

"Put everything on the center of the band?"

"That is probably the best frequency to use in a case of this kind."

"He can't control, so we'll push him down close to the ground before we go to work on him—so we don't have so far to fall if anything goes screwy with the works. Here's hoping nothing gives away!"

The *Sirius*, almost against the flaming screens of the Jovian, and both vessels very close to the surface of the satellite, Brandon tested the power leads briefly, adjusted dials and coils, then touched the button which actuated the relays—relays which in turn drove home the gigantic switches that launched a fearsome and as yet untried weapon. Instantly released, the full seven hundred thousand kilofranks of their stupendous batteries of accumulators drove into the middle frequency of the attacking band, and Brandon's heart was in his mouth as he stared into the plate to see what would happen. He saw! Everything in the *Sirius* held fast, and under the impact of the inconceivable plane of force, the screens of the enemy vessel flared instantly into an even more intense incandescence and in that same fleeting instant went down, and all defenses vanished as the metal sphere fell apart into two halves, as would an apple under the full blow of a broad-axe.

Brandon quickly shut off his power and stared in relief into the central compartment of the globular ship of space, now laid open, and saw there figures, one or two of which were moving weakly. As he looked, one of these feebly attempted to raise a peculiar, tubular something toward a helplessly fettered body. Even as Brandon snatched away the threatening weapon with a beam of force, he recognized the captive.

"Great Cat, there's Breckenridge!" he gasped, and directed a lifting beam upon the bound and unconscious prisoner. Rapidly, but carefully, he was brought through the double airlock and into the control room, where his shackles were cut away and where he soon revived under vigorous and skilful treatment.

"Any more of you in there? Did I hit any of you with that beam?" demanded Brandon, intensely, as soon as Breckenridge showed signs of understanding.

"King's in there somewhere, and there's a Callistonian human being that you mustn't kill," the chief pilot replied, weakly and with great effort in every word. "Don't believe that you hit anybody direct, but the shock was pretty bad." Having delivered his message, he lay back, exhausted.

"All x. Crown, give me a squad...."

"Not on your life!" barked the general. "This is my job and I'll do it myself. Your job is fighting the *Sirius*—stay with it!"

"Not in seven thousand years—I'm in on this, too," Brandon protested, but was decisively overruled by Newton.

"You belong right here at this board, since no one else can handle it the way you can. Stay here!" he commanded.

"All right," grudgingly assented the physicist, and held the *Sirius* upright, with her needle-sharp stern buried a few feet deep in the ground.

He watched the wreckage jealously while Crowninshield and forty helmeted men issued from the service door in the lower ultra-light compartment and advanced upon the two halves of the enemy vessel. As no hostile demonstrations ensued, scaling ladders were quickly placed and with weapons at the alert the police boarded the hemispheres, manacled the still helpless beings visible, and, after laying down a fog of stupefying gas, vanished into compartments beyond the metal partitions. After a short time they reappeared and climbed down the scaling ladders, carrying several inert forms, and Brandon spoke into his transmitter.

"King all x, Crowninshield?"

"I think so. Not being in the control room he was not as badly shocked by the passage of the beam as were Breckenridge and those you saw. The things in the other rooms were about ready to fight, so we gave them a little whiff of tritylamin, but Captain King will be as good as ever in a few minutes."

"Fine business!" The police entered the *Sirius*, the service doors clanged shut, and Brandon turned to Westfall.

"While they're coming up, I guess I'll pick up Perce and Miss Newton. We'd better get them aboard and beat it, while we're all in one piece!"

But even before he could send out the exploring beam of his communicator, the voice of Stevens came from the receiver.

"Hi, Brandon and Westfall! We've watched the whole show. Congratulations, fellows! Welcome to Ganymede! You are in our valley—we're upstream from you about three hundred meters; just below the falls, on the meadow side."

"All x," Brandon acknowledged. "We saw you. Come on out where we can pick you up. We've got to get away from here, and get away fast!"

"We'll carry off the pieces of that ship, too, Quince—we may be able to get a lot of pointers from it," and Brandon swung mighty tractor beams upon the severed halves of the Jovian vessel, then extended a couple of smaller rays to meet the two little figures racing across the smooth green meadow toward the *Sirius*.

CHAPTER X
Among Friends at Last

THE time for the landing of the *Sirius* was drawing near, and the castaways upon Ganymede had donned their only suits of earthly clothing, instead of the makeshifts of mole-skin, canvas, and leather they had been wearing so long. Thorns and underbrush had pierced and torn their once natty outing costumes, and sparks and flying drops of molten metal from Stevens' first crude forges had burned in them many gaping holes.

"I did the best I could with them, Steve, but they look pretty crumby," Nadia wrinkled her nose as she studied the anything but invisible seams, darns, and staring patches everywhere so evident, both in her own apparel of gray silk and in the heavy whipcord clothing of her companion.

"You did a great job, considering what you had to work with," he reassured her. "Besides, who cares about a few patches? I feel a lot more civilized in my own clothes, don't you?"

"Well ... yes," she admitted. "They're silk, anyway, even if they don't look like much, and I'm just reveling in the feel of them next to me after the horrible, rough, scratchy things I've been wearing. See anything yet?"

"Not yet." Stevens had been scanning the heavens with a pair of binoculars. "That doesn't mean much, though, as they'll be just about in the sun and they'll be coming like a scared dog. Might as well put away these glasses—we probably won't be able to see them until they're right on top of us."

"What shall we take with us?"

"Don't know—nothing, probably, since they must have a campaign already mapped out. I'd like to salvage a lot of this junk, but I'm afraid we won't be able to. I'm going to take my bow and arrows, though, aren't you?"

"Absolutely! That's one thing that's better than anything I ever had on Earth. This bow of mine is perfect."

"There they are! Three rousing cheers! Say, but that old hulk looks good to me!"

"Doesn't she, though!" cried Nadia, vibrant with excitement. "You know, Steve. I've hardly dared really to believe it until this very minute. Oh look! What's that?"

The *Sirius* had stopped in midair and they could see, far in the distance, the tiny sphere of the Jovians, rushing to the attack.

"Oh, how horrible!" cried the girl, her voice breaking. "I'm afraid, Steve...."

"You needn't be, ace. I've told you they won't go off half-cocked as long as Westfall is on the job. They're ready for anything, or they wouldn't be here—but just the same I wish that they had that Titanian mirror and a couple of those bombs!"

In a moment more the Jovian plane of force was launched, the powerful ray-screens flared into white-hot, sparkling defense, and the battle was on. Held spell-bound as the castaways were by that spectacular duel, yet Stevens' trained mind warned him of the perils of their position.

"Grab your bow and we'll beat it!" and he rapidly led her away from the steel structures to an open hillside, well away from any projection, tree, or sharp point of rock. "If that keeps up very long, we're going to see some real fireworks, and I don't know whether there will be enough left of our plant here to salvage or not. Everything is grounded, of course, but I don't believe that ordinary grounds will amount to much against what's coming."

"What *are* you talking about?" demanded Nadia.

"Look!" he replied, pointing, and as he spoke, a terrific bolt of lightning launched itself from the incandescent screen of the Jovian vessel upon their slender ultra-radio tower, which subsided instantly into a confused mass of molten and twisted metal.

AS the power of the beams was increased and as the combatants drew nearer and nearer the ground, the lightning display grew ever more violent. Well below the canyon as the warring vessels were, the power-plant and penstock did not suffer at all and only a few discharges struck the *Forlorn Hope*—discharges which were carried easily to ground by the enormous thickness of her armor—but every prominent object for hundreds of yards below the *Hope* was literally blasted out of existence. Radio tower, directors and fittings; trees, shrubs, sharp points of rock—all were struck again and again; fused, destroyed, utterly obliterated by the inconceivable energy being dissipated by those impregnable screens of force. Even almost flat upon the ground as the spectators were, each individual hair upon their heads strove fiercely to stand erect, so heavily charged was the very air. Stevens' arm was blue for days, such was Nadia's grip upon it, and she herself could scarcely breathe in that mighty arm's constriction—but each was conscious only of that incredibly violent struggle, of that duel to the death being waged there before their eyes with those frightful weapons, hitherto unknown to man. They saw the *Sirius* triumphant, and Stevens led the dancing girl back into their dwelling of steel.

"Danger's all over now. Radio's gone, but we should fret a lot about that. It has done its stuff—we can use the communicators. And now, sweetheart, I'm going to kiss you—for the first time in seven lifetimes."

Locked in each other's arms, they watched the scene until Stevens thought it time to send his message. Then, running hand in hand toward the huge space-cruiser, they were snatched apart and drawn up toward the double airlocks of the main entrance. Pressure gradually brought up to normal, they were ushered into the control room, where Nadia glanced around quickly and almost took her father off his feet by her tempestuous rush into his arms.

"Oh, Daddy darling. I just knew you'd come along! I haven't seen you for a million years!" she exclaimed, rapturously. "And Bill, too—wonderful!" as she fervently embraced a young man wearing the uniform of a lieutenant of Interplanetary Police. "Ouch, Bill—you're breaking all my ribs!"

"Well, you cracked three of mine. Maybe you don't know how husky you are, but you've got a squeeze like a full grown boa constrictor!" He held her off at arms' length and studied her with admiration. "Gee, it's fine to see you again, Sis. You're looking great, too—I think I'll bring my girl out here to live. You always were a knockout, but now you're the loveliest thing I ever saw!"

He made his way through the group surrounding Stevens, while Nadia and her father talked earnestly.

"I'm Bill Newton. Thanks," he said, simply, holding out his hand, which was taken in a bone-crushing grip.

"Bring him over here, Bill!" Nadia called before Stevens could find a reply.

"I don't know how to say anything, Stevens," the officer continued, in embarrassment, as the two men turned to obey the summons. "She's a good kid, and we think a lot of her. We'd about given her up. We.... She.... Oh, rats, what's the use? You know what I mean. You're there, Stevens, like a...."

"Clam it, ace!" Stevens interrupted. "I get you, to nineteen decimals. And you don't half know just what a good kid she really is. She's the reason we're here—we were down pretty close to bed-rock for a while, she stood up when I wilted. She's got everything. She...."

"Clam it yourself, Steve! Don't believe a word of it, Dad and Bill. *Wilt*!" Nadia's voice dripped scorn. "Why, he di...."

"Please!" Newton's voice was somewhat husky as he silenced the clamor of the three young people, all talking at once. "I will not embarrass you further by trying to say something that no words can express. You told me that you would take care of her, and I learn that you have done so."

"I did what I could, but most of the credit belongs to her, no matter what she says," Stevens insisted. "Anyway, sir, here she is; alive, well and ... unharmed," and his eyes bore unflinchingly the piercing gaze of the older man, who was reassured and pleased by what he read therein. "One thing I want to say right now, though, that may make you feel like canceling the welcome. I loved Nadia even before the *Arcturus* was attacked, and since then, coming to know her as I have, the feeling hasn't lessened any."

"Nadia has already told me all about you two," said her father, "and the welcome stands. If you could take care of her as well as you have done since you left the *Arcturus*, I have no doubt of your ability to take care of her for life. We have been examining the work you have done here, son, and the more I saw of it the more amazed I became that you could have succeeded as you did. We are deeply indebted.... Just a minute! There's my call—I'm wanted in Fifteen. I'll see you again directly."

"Hi, Norm!" Stevens further relieved the surcharged atmosphere. "As soon as you and Quince can leave those controls come over and see us, will you?"

"All x—coming up!" sounded Brandon's deep and pleasant bass, and the two rescuers, who had tactfully avoided the family reunion, came over and greeted the third of their triumvirate.

"Ho, Perce—you look fit." Brandon ran an expert hand over Stevens' arm and shoulder. "Looks as if he might last a round or two, doesn't he, Quince?"

"You are looking fine, Steve. Neither of you appear any the worse for your experiences. So this is Nadia? We have heard of you, Miss Newton."

"I believe that, knowing Dad," she replied. "Thanks, both of you, for digging us out. I've heard about you two, and I'm going to kiss you both."

Westfall, the silent and reserved, was taken aback, but Brandon met her more than half-way.

"All x, Nadia—payment in full received and hereby acknowledged," he laughed, as he allowed her feet to return to the floor. "Even if it was some stout lads from Mars and Venus that did all the work we'll take the reward—especially since Alcantro and Fedanzo couldn't feel even such a high-voltage salute as that one was, and I can't picture you kissing a Venerian even if you could get to him. Whenever you get lost again, be sure to let us know, now that you've got our address. If I know Perce at all, you've heard of us 'til you're sick of it and us—it's a weakness of his—talking too much."

"Why, it's no such th…." began Nadia, but broke off as an aide came up and saluted smartly.

"Pardon me, but General Crowninshield requests that Doctor Brandon, Doctor Westfall, and Doctor Stevens join the council in Lounge Fifteen as soon as convenient." He saluted again and turned away.

"Yes, that's right, folks—we've got to take a lot of steps, fast—see you later," and Brandon, taking each of the other two by an arm, marched them away toward the designated assembly room.

THERE, already seated at a long table, were Czuv, King, and Breckenridge, all fully recovered, engaged in earnest conversation with Newton and Crowninshield. Alcantro and Fedanzo, the Martian scientists, were listening intently, as were the two Venerians Dol Kenor and Pyraz Amonar. The eyes of the three newcomers, however, did not linger upon the group at the table, but were irresistibly drawn to one corner of the room, where six creatures lay in the heaviest manacles afforded by the stores of the Interplanetary Police. Not only were they manacled, but each was facing a ray-projector, held by a soldier whose expression showed plainly that he would rather press the lethal contact than not.

"Oh—those the things we're fighting?" Brandon stopped at the threshold and stared intently at the captive hexans. Goggling green eyes glaring venomously, they were lying quiet, but tense; mighty muscles ready to burst into berserk activity should the attention of a guard waver for a single instant.

But little more than half as large as the savage creatures with whom Stevens had fought in the mountain glade upon Ganymede, the hexans resembled those aborigines only as civilized men might resemble gigantic primordial savages of our own Earth. Brandon's gaze went from short, powerful legs up a round, red body to the enormous, freakish double pair of shoulders, with its peculiar universal jointing. From the double shoulders sprang four limbs, the front pair of which were undoubtedly arms, terminating in large, but fairly normal, hands. The intermediate limbs were longer than the legs and were much more powerful than the arms, and ended in members that were very evidently feet and hands combined. What in a human being would be the back of the hand was the sole of the foot—when walking upon that foot the long and dexterous thumb and fingers were curled up, out of the way and protected from injury, in the palm of the hand. From the monstrous shoulders there rose a rather long and very flexible, yet massive and columnar neck, supporting a head neither human nor bestial—a head utterly unknown to Terrestrial history or experience. The massive cranium bespoke a highly developed and intelligent brain, as did the three large and expressive, peculiar, triangular eyes. The three sensitive ears were very long, erect, and sharply pointed. Each was set immediately above an eye, one upon each side of the head and one in front. Each ear was independently and instantly movable in any direction, to catch the faintest sound. The head, like the body and limbs, was entirely devoid of hair. The horns, so prominent in the savages Stevens had seen, were in this highly intelligent race but vestigial—three small, sharp, black protuberances only an inch in length, one surmounting each ear, outlining the lofty forehead. The nose occupied almost the whole middle of the face and was not really a nose—it developed into a small and active proboscis. The chin was receding almost to the point of disappearance, so that the mouth, with its multiple rows of small, sharp, gleaming-white teeth, was almost hidden under the face instead of being a part of it. Such were the hexans, at whom the Big Three stared in undisguised amazement.

"Attention, please!" Newton called the meeting to order. "We have learned that all the passengers of the *Arcturus*, and all the crew save three, are alive and safe for the time being. Most of them are upon the satellite Europa. However, I understand that we are not yet sufficiently well armed to withstand such an attack in force as will certainly develop when we move to rescue them. This seems to be a war of applied physics—Doctor Brandon, as spokesman for the Scientific forces of the expedition, what are your suggestions?"

"Anticipating an attack in response to signals probably sent out by the enemy," replied Brandon. "I headed directly south immediately. We are now well south the ecliptic, and are traveling at considerably more than full Martian acceleration. Before making any suggestions, I should like to hear from Captain Czuv, who is more familiar than we are with the common enemy. Are they apt to follow us: can they detect us if we should drift at constant velocity; and can we search the brains of the prisoners with his Callistonian thought-exchanger, if he should build one with our help?"

"If they are close enough to us to overtake us without too much lost time, they will certainly attack us," Czuv answered at a nod from Newton. "Ordinarily they would pursue us to the limits of the Solar System if necessary, but since they have suffered reverses of late and cannot spare any vessels, they will probably not pursue us far. Yes, they can detect us, even without the driving rays, since this vessel uses much low-tension, low-frequency electricity in its automatic machinery, lights, and so on. No; our thought-transformer cannot take thoughts by force, and the hexans will exchange no ideas with us. They are implacable and deadly foes of all humanity, irrespective of planet or race. Mercy is to them unknown—they neither give nor take quarter."

"I can bear him out in that," Crowninshield interposed grimly. "The first one to recover snapped our ordinary handcuffs like so much thread and literally tore four men to pieces before the rest of us could ray him. Will you need me longer, Director Newton?"

"I think not. General. Captain Czuv, you have made no headway with them?" asked the Director.

"None whatever, as I foretold. They understand me thoroughly, since two of them speak my own tongue, but nothing that they have said can ever be repeated here. I knew from the first that all such attempts would be fruitless, but I have tried—and failed. I suggest what I suggested at first—put them to death, here and now, as they lie there, for most assuredly they will in some way contrive to take toll of lives of your own humanity if you allow them to live."

"You may be right," said Newton, "but neither the General nor myself can give the order for their death, since Interplanetary law does not countenance such summary action. However, the guards are fully warned of the peril, and will ray every prisoner at the first sign of unruliness. General Crowninshield, you may remove the prisoners and deal with them in accordance with...."

PANDEMONIUM reigned. At Crowninshield's signal for the guards to leave the room with their captives, all six had strained furiously at their bonds and three of them had broken free in a flash, throwing themselves upon the guards with unthinkable ferocity. Stevens, seeing a ray-projector in a hand of one of the prisoners, hurled his heavy chair instantly and with terrific force. The projector flew into the air, shattered and useless, while the hexan was knocked into a corner by the momentum of the massive projectile and lay there, stunned and broken. Brandon, likewise reacting instantaneously, had bent over and seized a leg of the table, bracing his knee against the corner. With a mighty lunge of his powerful body he wrenched out the support and with a continuation of the same motion, he brought the jagged oak head of his terrible club down full upon the crown of the second hexan, who had already torn one guard apart and was leaping toward Czuv, his hereditary foe. In midflight he was dashed to the floor, his head a shapeless, pulpy mass, and Brandon, bludgeon again aloft, strode deeper into the fray. For a brief moment searing lethal beams probed here and there, chains clanked and snapped, once more that ponderous and irresistible oaken mace fell like the hammer of Thor, again spattering brains and blood abroad as it descended—then again came silence. The six erstwhile prisoners lay dead, but they had taken five of the guards with them—literally dismembered, hideously torn limb from limb by the superhuman, incredible physical strength and utter ferocity of the hexans.

By common consent the meeting was adjourned to another room, for the business in hand could not be postponed.

"Captain Czuv was right—we Tellurians could not believe in the existence of such a race without the evidence of our own senses." Newton reopened the meeting. "From this time on we take no prisoners. Doctor Brandon, you may resume."

"The detectors and lookouts will give ample warning of any attack, and Doctor Westfall has suggested that we should have all possible facts at hand before we try to decide upon a course of action. We should like to hear the full reports of Captain King, Captain Czuv, Chief Pilot Breckenridge, and Doctor Stevens."

The four men told their stories tersely and rapidly, while the others listened in deep attention. As the last speaker sat down, Newton again turned to Brandon, who silently jerked his head at Westfall, knowing his own inadequacy in such a situation—realizing that here was needed Westfall's cold and methodical thinking.

"Director Newton and gentlemen," Westfall spoke calmly and precisely. "We have much to do before we can meet the hexans upon equal terms. We have many new fields of force and rays to develop, of whose nature and necessity Doctor Brandon is already aware. Then, too, we must recalculate our visirays so that we can operate at greater range and efficiency. We must also examine the hexan space-ship which is towing, to do which it will be desirable to drift at constant velocity for a time. In it we may find instruments or devices as yet unknown to us. It also occurs to me that since this is an Interplanetary Police problem of the first magnitude, we should at once get in touch with Police Headquarters, so that the Peace Fleet can be armed as we ourselves are, or shall be, armed; for a large and highly efficient fleet will be necessary to do that which must be done. It is, of course, a foregone conclusion that Interplanetary humanity will support the humanity of Callisto against the hexans.

"It is also self-evident that we must stay here and rescue the Tellurians now upon Europa and Callisto, but we are not yet in position to decide just how that rescue is to be accomplished. Four courses are apparently open to us. First, to attempt it as soon as we shall have strengthened our armament as much as is now possible. That would invite a massed attack, and in my opinion would be foolish—probably suicidal. Second, to stand by at a distance until the rocket-ship is launched, then to escort it back to the Earth. Third, to aid the Callistonians as much as possible while awaiting the completion of the rocket-vessel. Fourth, and perhaps the most feasible and quickest, it may be possible for the Callistonian rocket-ships to bring out fellow-Tellurians, a few at a time, to us here out in space, since they are apparently able to come and go at will. However, I would recommend that we make no plans for the rescue as yet—there is little use in attempting to deal with an ever-changing situation until we are ready to act forthwith. I suggest that we strengthen our offensive and defensive armament first, then secure information as to the exact status of affairs, both upon Callisto and upon Europa. Then, ready to act, we will do at once whatever seems called for by the situation then obtaining."

"The program as outlined seems eminently sensible. Are there any comments or suggestions?" None having been offered, Director Newton adjourned the meeting and each man attacked his particular problem.

True to Czuv's prediction the hexans did not deem it worthwhile to pursue the Terrestrial vessel, so obviously and so earnestly fleeing from them, and shortly, the acceleration was cut off, to render possible a thorough study of the two halves of the spherical warship of the enemy. Scientists donned space-suits and studied every feature of the strange vessel, while mechanics dismantled and transferred to the *Sirius* every device and instrument of interest. One or two novel and useful applications of rays and forces were found, their visirays and communicators in particular being of a high degree of efficiency; but upon the whole the science of the hexans was found to be inferior to that now known to the scientists of Interplanetary's flying laboratory. Brandon studied the hexan power-system most carefully, and, everything in readiness and after a long talk with Westfall, he called a general conference in the control-room.

"Gentlemen, we have done about everything we can do for the time being. By combining the best features of the visirays and communicators of the hexans with our own newly-perfected devices, we now have a really excellent system of communication. Our friends from Mars and Venus have so altered and enlarged our force-controls that our offensive and defensive fields, rays, and screens leave little to be desired. In power we are far ahead of the enemy. They apparently know nothing of the possibilities of cosmic radiation, but depend upon tight-beam transmission from their own power-plants—which transmission they have perfected to a point far beyond anything reached by us of the three planets. They do not use accumulators, and therefore their dissipation is limited to their maximum reception, which is about seventy thousand kilofranks. Since we can dissipate ten times that amount of energy, we could withstand, for a short time, the simultaneous attacks of ten of their vessels. Eleven or more of them, however, would be able to crush our defensive screens—and Captain Czuv has seen as many as a hundred of their space-ships in one formation. Furthermore, since they have several times our maximum acceleration, they could concentrate quickly upon any desired point. We could not escape them by flight if they really set out to overtake us, which they certainly will do if we again venture into their territory. Therefore it is clear that we cannot subject ourselves to any attack in force and it follows that we cannot do much of anything until the police fleet of some five hundred vessels can be re-armed and can join us near Callisto. This will require several months at best. As you already know, it has been decided that we should not return to any of the minor planets, as to do so might invite a hexan attack upon our police fleet which is as yet unprepared. We are now heading for Uranus, in the hope that such a course will distract the attention of the hexans from Tellus, even though they probably already know that we are Tellurians. Our new communicator ray will reach any member of the Jovian system from this point. It has been decided that it is safe to use it, since it employs an almost absolutely tight beam of very small diameter, and since we know that that one hexan vessel, at least, had no apparatus sufficiently sensitive to detect a beam of that nature. We will therefore now get in touch with the Callistonians and with our own people."

BRANDON seated himself before the communicator screen, and while the others packed themselves closely around his stool, he snapped on the visiray and turned the dials which directed that invisible, immensely complex beam through space. The screen was apparently in itself a coign of vantage, flying through space with the velocity of light, and the watchers gasped involuntarily and drew themselves together, as with that unthinkable speed they flashed down toward the surface of Callisto. So realistic was the impression that they themselves were hurtling through the void, that they could scarcely reason themselves into believing their positive knowledge that the impending collision was not an actual happening! Reducing the velocity of the projection abruptly as it approached the satellite, Brandon flashed it down into a crater indicated by Czuv, and along a tunnel to the city of Zbardk, where the Callistonian captain held a long conversation with the Council of the nation. Frowning in thought, he turned to Newton and spoke seriously and slowly.

"Immediately after the loss of our super-plane, with the supposed death of King, Breckenridge, and myself, the other Tellurian officers were returned to Europa, since even they could be of no assistance to us Callistonians in our struggle against the new, high-acceleration vessels of the hexans. The present situation is much more serious than I would have believed possible. The last vessel going to visit Wruszk, our city upon Europa, was caught and destroyed by the hexans, and for many weeks no ship or message has come from there to Callisto. In spite of the fact that the hexan fleet is smaller than ever before, they are guarding Europa very closely. It is feared that they may have found and destroyed our city there—an expedition is even now about to set out in a desperate attempt to learn the fate of our fellows."

"Suppose the rays of the lifeboats were detected in landing?" asked Brandon. "That might have given them a clue."

"Possibly; but it is equally possible that our own men became careless in the operation of one of our own vessels. Having been unmolested so long, they might have relaxed their vigilance. We may never know."

"Tell 'em to cancel the expedition—we'll shoot the visiray over there right now and find out all about it. We'll let them know pretty quickly. Also, you might tell them that you've got complete plans and specifications for all the weapons that the hexans have, and a couple besides, and that the quicker they shoot a ship out here after you, the sooner they can get to building some stuff to blow those hexans clear out of space!"

It was the work of only a few moments to drive the visiray projection to Europa, where Czuv, to the great relief of all, found that the hexans had not yet discovered either Wruszk or the Terrestrial workings. All Europan humanity, fully aware of the hexan investment, was exerting every possible precaution against discovery by the enemy. This information was duly flashed to the Council of Callisto, and the projection was then hurled across the intervening reaches of space and into the cavern in which was being built the enormous rocket-ship in which the Terrestrial refugees were to attempt the long voyage back to their own distant planet.

It took some little time to convince Doctor Penfield that there had been projected into the empty air of his little sanctum an absolutely invisible and impalpable structure of pure force capable of receiving and transmitting voice and vision. Once convinced of the reality of the phenomenon, however, the speaker beside Brandon's communicator screen fairly rattled under the fervor of his greeting, so great was his pleasure at the arrival of the expedition of relief and in knowing that King and Breckenridge, whom they had, of course, given up for dead, were aboard the Interplanetary vessel.

Penfield reported that the work upon the great rocket-ship was progressing satisfactorily, although slowly, since it was so much larger than any vessel theretofore constructed by the Callistonians. Newton, in turn, informed the autocrat of the stranded Terrestrials as to the *status quo* of the rescuing party.

"Of course, because of the hexan blockade, you cannot take us off until they have been wiped out, which will be several months at best," the surgeon said, slowly, and a shadow came over his face as he spoke. "Well, what can't be cured...."

"Trouble with the personnel?" King broke in sharply.

"Personnel, yes; but not trouble in the sense you mean—we have had none of that. It is only that there are four more of us now than there were...."

"Huh? How come?" demanded Brandon, in astonishment.

"Four babies have been born to us here so far, and several more are coming. They are the ones I'm worried about. Most normal adults can stand it here without any serious effects, but this thin atmosphere and weak gravity are certain to result in abnormal development of children. However, there may be another way out of it. Are you using normal acceleration, or have you Martians aboard?"

"Both," replied Brandon. "We are carrying two inhabitants of Mars, but Alcantro and Fedanzo are not ordinary Martians. They have been in constant training ever since we left Tellus, and now they can stand as high an acceleration as a weak Tellurian. We're riding at normal."

"Good! As you already know, there has been no communication of late between here and Callisto. It had already been decided, however, that one more voyage must be risked, in order to bring back material which is most urgently needed. Since the vessel will leave here light and is large enough to carry about thirty passengers on a short trip with some crowding, the Council will probably approve of having it carry some of our passengers out to the *Sirius*—especially now, since a vessel must visit you, anyway, to get Captain Czuv and the specifications of the new armament. All these things can be done with one vessel in one trip."

"That sounds fine!" boomed King. "It will give me a chance to get back there where I belong, too. Whom are you sending out?"

"The seven couples who either have babies already or who will have them in the next few months; and some of our young who aren't standing the gaff any too well. You won't be in the red very deeply on the deal, either—while two or three of the passengers I am sending you will certainly be a nuisance; anybody could use, anywhere, such men as Commander Sanderson and Lieut..."

"Sanderson!" interrupted King. "Why, he wasn't—when did *he* get married?"

"The day after we arrived here," replied the surgeon. "His fiancee was aboard the *Arcturus*, and when they found out how long we would have to be here, they very sensibly decided not to wait."

"Were there any others?" demanded Nadia, who, standing between Stevens and her father, had been an interested listener.

"Plenty of them! Fourteen of our young women passengers have married here upon Europa. A few married fellow-passengers, but most of them picked out officers of the *Arcturus*. You'll find your staff made up pretty largely of benedicts now, King! We've been here a year, you know, and time will tell! Young Commander Sanderson's a fine baby—he'll be a credit to the IPC some day, if we can get him aboard the *Sirius*, where he can get a good start. We could give our babies normal air pressure here by building special rooms, but we cannot give them the normal acceleration necessary to develop their muscles properly."

"Well, we'd better snap over to Callisto and take this up with the Council," Brandon put in. "I don't imagine that there will be any objections, so you might as well get your ship gassed up and loaded—we'll be back here with the okay in about a minute and a half."

WITH Brandon at the controls and with Czuv at the communicator plate, the projection flashed toward distant Callisto and the group melted away, each man going about his interrupted task.

"Daddy, take us somewhere—I want to talk to you," Nadia spoke to her father, and the director led her and Stevens to his own room.

"All x, daughter; out with it!" and he bent upon her a quizzical glance, under which a fiery blush burned from her throat to her forehead.

"Dad, I've been thinking a lot since you rescued us, and what we've just heard has given me the nerve to say it. Steve, of course, wouldn't dare suggest such a thing until we're safely back on Earth, so I will." Her deep brown eyes held his steadily. "All those girls got married—why, some of them have babies already—and Steve and I have waited for each other *so* long, daddy! And *none* of them love each other the way we do. Do they, Steve?"

"I don't see how they could, sir; and that goes straight across the panel," and he bore unflinchingly the piercing gaze of the older man as his right arm encircled the girl and held her close.

"Well, why not?" A sudden smile transformed Newton's stern visage. "There are three chaplains with the police—a Methodist minister, a Catholic priest, and a Jewish rabbi. Also, we have on board two full-fledged I-P captains, either of whom is authorized to tie matrimonial knots. The means are not lacking—if you're both sure of yourselves?" and all levity disappeared as he studied the two young faces.

"Yes, you are sure," he continued after a moment "just as her mother and I were—and are. It is too bad that she cannot be here with you, but it may be a long time before we can return to Tellus, and you have indeed waited long."

"Oh thanks, Daddy, you're just a perfectly wonderful old darling!" Nadia exclaimed, as she threw her arms rapturously around his neck. "And this isn't a warship at all—you know perfectly well that it's a research laboratory, and that as soon as the Navy gets here, you won't let it fight a bit more, because such scientists can't be allowed to risk themselves! And also, you're forgetting that whole flock of women and babies that are coming out here just as fast as they can get themselves ready. So get going, daddy old dear, and let's do things! Steve's a Quaker and we're Presbyterians, so none of the chaplains will do at all. Besides, I promised Captain King ages ago that he could marry me, so go get him and we'll do it now. Bill can be my bridesmaid, you'll give me away, and Steve can have the other two of his Big Three for best men. I'm off to hunt up the flimsiest, fussiest white dress I can find in my trunks. Let's go!"

"Mr. Newton." Stevens spoke thoughtfully as Nadia darted away. "You said something about her mother, I didn't want to say anything to raise false hopes while she was here, but I've got an idea. Let's meet in Brandon's room instead of here. We can send code to Tellus easily enough on our ultrawave, and we may be able to fake up something on vision."

A few minutes later the Big Three were in Brandon's private study; staring intently into a screen of ground glass upon which played flickering, flashing lights, while the black-haired physicist manipulated micrometer dials in infinitesimal arcs.

"Once more, Mac," Brandon directed. "Pretty nearly had them that time. We're stretching this projector about six hundred percent, but we've got to make this connection. Can't you give me just a little more voltage on those secondaries?"

"I can *not*!" the voice of the first assistant snapped from the speaker. "I'm overloading now so badly that some of my plates are getting hot—if I hold this voltage much longer, the whole secondary bank of tubes is going out. All x—you're on zero!"

"All x!" Flashing and waning, the lights upon the screen formed fleeting, shifting, nebulous images of a relay station upon distant Earth; but the utmost power of the transmitting fields could neither steady the image nor hold it.

"Back off, Mac," Brandon instructed. "I'm afraid we can't hold 'em direct—no use blowing a bank of tubes. We'll try relaying through Mars—we can hold them there, I think. It will muss up reception some, but it will probably be better than direct, at that. Point oh five three six ... all x—shoot!"

Brandon's relay station upon Mars was finally raised and held, and a corps of keenly interested engineers there made short work of the Earth-Mars linkage. Soon the screen glowed with the picture of the transmitter-room of the Terrestrial station, and while the three men were waiting for Mrs. Newton to be called to her own television set, the door behind them opened. Nadia and her escorts entered the room—but Stevens' eyes saw only the entrancing vision of loveliness that was his bride. Dressed in a clinging white gown of shimmering silk, her hair a golden blond corona, sweetly curved lips slightly parted and wide eyes eloquent, she paused momentarily as Stevens came to his feet and stared at her, his very heart in his eyes.

"You never saw me in a dress before—do you like me, Steve?"

"*Like* you! You're beautiful!" and gray eyes and brown, deep with wonder and with love, met and held as, unheeding the presence of their friends, they went into each other's arms in a coalescence as inevitable and as final as Fate itself.

"Hi, Nadia old dear!" and "Daughter, from what I can see of my son-in-law, I believe that he may do," came together from the speaker. Nadia tore herself from Stevens' embrace, to see upon the lambent screen the happily smiling faces of her mother and sister.

"Mother! Claire! Oh, you three wonder-workers!" She addressed simultaneously the distant Terrestrials and the scientists at her side, while broken exclamations, punctuated by ominous, crackling snaps, came from the laboring amplifier.

"Sorry to interrupt," MacDonald's voice broke in, "but you'll have to hurry it up. Alcantro and Fedanzo are doing their best, but every plate in my secondary bank's red hot, and you could fry an egg on any one of my transformers. Even my primary tubes are running hot. She won't hold together five minutes longer!"

Captain King opened his book, and in that small steel room, unadorned save for stack upon stack of bookcases, the brief but solemn ceremony joining two young lives was read—its solemnity only intensified by its unique accompaniment. For from Brandon at the primary controls, through the power-room of the *Sirius* and the relay-station upon Mars, to the immense Interplanetary transmitter upon Earth, the greatest radio and television engineers of two planets were fighting overdriven equipment, trying to hold an almost impossible connection, in order that Nadia Newton's mother and sister might be present at her wedding, hundreds of millions of miles distant in space!

"I pronounce you man and wife. Whom God hath joined, let no man put asunder." The sacred old ritual ended and Captain King picked up the bride in his great arms as though she were a baby, kissed her vigorously, and set her down in front of the transmitter. In the midst of the joyous confusion that ensued a tearing, rattling crash came from the speaker and the screen went blank.

"There!" lamented MacDonald from the power room. "I knew they'd blow! There goes my whole secondary bank—eight perfectly good ten-nineteens all shot to...."

"That's too bad, but it couldn't be helped; they went for a good cause," interrupted Brandon. "I'll come down and help clean up the mess."

LEAVING the bridal party, he made his way rapidly to the power room, where he found MacDonald and the two Martians inspecting the smoking remains of what had been the secondary bank of their powerful ultra-transmitter. Spare parts in abundance were on hand, and it was not long until the damaged section was apparently as good as new.

"Now to try her out," Brandon announced. "We want to give her a good workout, but there's no use trying the I-P stations any more—they're altogether too hard to handle at this range. Czuv said something about an unknown race of monstrosities at the south pole of Jupiter—let's try it on them for a while."

He flung the field of force out into space, as responsive to his will as a well-trained horse, and guided it toward the southern limb of that gigantic world. Down and down the projection plunged, through mile after mile of reeking, steaming fog, impenetrable to earthly eyes. Finally it came to rest upon the surface, hundreds of feet deep in a lush, dank, tropical jungle, and Brandon plugged into the Venerian room.

"Kenor? We've got a lot of use for you, if you can come down here for a while. Thanks a lot." He turned to the Martians. "Luckily, we've got a couple of infra-red transformers aboard, so we won't have to build one. You fellows might break one out and shunt it onto this circuit while Dol Kenor is hunting up something for us to look at.

"Hi, old Infra-Eyes!" he went on, as the Venerian scientist waddled into the room in his bulging space-suit. "We've got something here that's right down your alley. Want to see what you can see?"

"Ah, a beautiful scene!" exclaimed Dol Kenor, after one glance into the plate. "It is indeed a relief, after all this coldness and glare, to see such a soft, warm landscape—even though I have never expected to behold quite such a violent bit of jungle," and under his guidance the projection flashed over hundreds of miles of territory. To the eyes of the Terrestrials the screen revealed only a blank, amorphous grayness, through which at times there shot lines and masses of vague and meaningless form; but the Venerian was very evidently seeing and enjoying many and diverse scenes.

"There, I think, is what you wish to see first," he announced, as he finally steadied the controls, and Brandon cut in upon the shunting screen the infra-red transformer. This device, developed long before to render possible the use of Terrestrial eyes in the opaque atmosphere of Venus, stepped up the fog-piercing long waves into the frequencies of light capable of affecting the earthly retina. Instantly the dull gray blank of the shunting screen became transformed into a clear and colorful picture of the great city of the Jovians of the South.

"Great Cat!" Brandon exclaimed. "Flying fortresses is right! They're in war formation, too, or I'm a polyp! We've got to watch this, Mac, all of it, and watch it close—it's apt to have a big bearing on what we'll have to do, before they get done. Better we rig up another set, and put a relay of observers on this job!"

CHAPTER XI
The Vorkul-Hexan War

VORKULIA, the city of the Vorkuls, was an immense seven-pointed star. At its center, directly upon the south pole of Jupiter, rose a tremendous shaft—its cross-section likewise a tapering seven-pointed star—which housed the directing intelligence of the nation. Radiating from the seven cardinal points of the building were short lanes leading to star-shaped open plots, from which in turn branched out ways to other stellate areas; ways reaching, after many such steps, to the towering inner walls of the metropolis. The outer walls, still loftier and even more massive ramparts of sullen gray-green metal, formed a seamless, jointless barrier against an utterly indescribable foe; a barrier whose outer faces radiated constantly a searing, coruscating green emanation. Metal alone could not long have barred that voracious and implacably relentless enemy, but against that lethal green emanation even that ravening Jovian jungle could not prevail, but fell back, impotent. Writhing and crawling, loathesomely palpitant with an unspeakable exuberance of foul and repellent vigor, possible only to such meteorological conditions as obtained there, it threw its most hideously prolific growths against that radiant wall in vain.

The short, zig-zag lanes, the ways, and the seven-pointed areas were paved with a greenish glass. This pavement was intended solely to prevent vegetable growth and carried no traffic whatever, since few indeed of the Vorkuls have ever been earth-bound and all traffic was in the air. The principal purpose of the openings was to separate, and thus to render accessible by air, the mighty buildings which, level upon level, towered upward, with airships hovering at or anchored to doorways and entrances at every level. Buildings, entrances, everything visible—all replicated, reiterated, repeated infinite variations in the one theme, that of the septenate stelliform.

Color ran riot; masses varied from immense blocks of awe-inspiring grandeur to delicate tracery of sheerest gossamer; lights flamed and flared in wide bands and in narrow, flashing pencils—but in all, through all, over all, and dominating all was the Seven-Pointed Star.

In and almost filling the space, at least a mile in width, between the inner and the outer walls were huge, seven-sided structures—featureless, squat, forbidding heptagons of dull green metal. No thing living was to be seen in that space. Its pavement was of solid metal and immensely thick, and that metal, as well as that of the walls, was burned and blackened and seared as though by numberless exposures to intolerable flame. In a lower compartment of one of these enormous heptagons Vortel Kromodeor, First Projector Officer, rested before a gigantic and complex instrument board. He was at ease—his huge wings folded, his sinuous length coiled comfortably in slack loops about two horizontal bars. But at least one enormous, extensible eye was always pointed toward the board, always was at least one nimble and bat-like ear cocked attentively in the direction of the signal panel.

A whistling, shrieking ululation rent the air and the officer's coils tightened as he reared a few feet of his length upright, shooting out half a dozen tentacular arms to various switches and controls upon his board, while throughout the great heptagon, hundreds of other Vorkuls sprang to attention at their assigned posts of duty. As the howling wail came to a climax in a blast of sound Kromodeor threw over a lever, as did every other projector officer in every other heptagon, and there was made plain to any observer the reason for the burns and scars in the tortured space between the lofty inner and outer walls of Vorkulia. For these heptagons were the monstrous flying fortresses which Czuv had occasionally seen from afar, as they went upon some unusual errand above the Jovian banks of mist, and which Brandon was soon to see in his visiray screen. The seared and disfigured metal of the pavement and walls was made so by the release of the furious blasts of energy necessary to raise those untold thousands of tons of mass against the attraction of Jupiter, more than two and a half times the gravity of our own world! Vast volumes of flaming energy shrieked from the ports. Wave upon wave, flooding the heptagons, it dashed back and forth upon the heavy metal between the walls. As more and more of the inconceivable power of those Titanic generators was unleashed, it boiled forth in a devastating flood which, striking the walls, rebounded and leaped vertically far above even those mighty ramparts. Even the enormous thickness of the highly conducting metal could not absorb all the energy of that intolerable blast, and immediately beneath the ports new seven-pointed areas of disfigurement appeared as those terrific flying fortresses were finally wrenched from the ground and hurled upward.

HIGH in the air, another signal wailed up and down a peculiar scale of sound and the mighty host of vessels formed smoothly into symmetrical groups of seven. Each group then moved with mathematical precision into its allotted position in a complex geometrical formation—a gigantic, seven-ribbed, duplex cone in space. The flagship flew at the apex of this stupendous formation; behind, and protected by, the full power of the other floating citadels of the forty-nine groups of seven. Due north, the amazing armada sped in rigorous alignment, flying along a predetermined meridian—due north!

At the end of his watch Kromodeor relinquished his board to the officer relieving him and shot into the air, propelled by the straightening of the powerful coils of his snake-like body and tail. Wings half spread, lateral and vertical ruddering fins outthrust, he soared across the room toward a low opening. Just before they struck the wall upon either side of the doorway the great wings snapped shut, the fins retracted, and the long and heavy body struck the floor of the passage without a jar. With a wriggling, serpentine motion he sped like a vibrant arrow along the hall and into a wardroom. There, after a brief glance around the room, he coiled up beside a fellow officer who, with one eye, was negligently reading a scroll held in three or four hands; while with another eye, poised upon its slender pedicle, he watched a moving picture upon a television screen.

"Hello, Kromodeor," Wixill, Chief Power Officer [2] greeted the newcomer in the wailing, hissing language of the Vorkuls. He tossed the scroll into the air, where it instantly rolled into a tight cylinder and shot into an opening in the wall of the room. "Glad to see you. Books and shows are all right on practice cruises, but I can't seem to work up much enthusiasm about such things now."

Kromodeor elevated an eye and studied the screen, upon which, to the accompaniment of whistling, shrieking sound, whirled and gyrated an interlacing group of serpentine forms.

"A good show, Wixill," the projector officer replied, "but nothing to hold the attention of men engaged in what we are doing. Think of it! After twenty years of preparation—two long lifetimes—and for the first time in our history, we are actually going to war!"

"I have thought of it at length. It is disgusting. Compelled to traffic with an alien form of life! Were it not to end in the extinction of those unspeakable hexans, it would be futile to the point of silliness. I cannot understand them at all. There is ample room upon this planet for all of us. Our races combined are not using one seven-thousandth of its surface. You would think that they would shun all strangers. Yet for ages have they attacked us, refusing to let us alone, until finally they forced us to prepare means for their destruction. They seem as senselessly savage as the jungle growths, and, but for their very evident intelligence, one would class them as such. You would think that, being intelligent and being alien to us, they would not have anything to do with us in any way, peacefully or otherwise. However, their intrusions and depredations are about to end."

"They certainly are. Vorkulia has endured much—too much—but I am glad that our forefathers did not decide to exterminate them sooner. If they had, we could not have been doing this now."

"There speaks the rashness of youth, Kromodeor. It is a violation of all our instincts to have any commerce with outsiders, as you will learn as soon as you see one of them. Then, too, we will lose heavily. Since we have studied their armaments so long, and have subjected every phase of the situation to statistical analysis, it is certain that we are to succeed—but you also know at what cost."

"Two-sevenths of our force, with a probable error of one in seven," replied the younger Vorkul. "And because that figure cannot be improved within the next seven years and because of the exceptional weakness of the hexans due to their unexpectedly great losses upon Callisto, we are attacking at this time. Their spherical vessels are nothing, of course. It is in the reduction of the city that we will lose men and vessels. But at that, each of us has five chances in seven of returning, which is good enough odds—much better than we had in that last expedition into the jungle. But by the Mighty Seven, I shall make myself wrap around one hexan, for my brother's sake," and his coils tightened unconsciously. "Hideous, repulsive monstrosities! Creatures so horrible should not be allowed to live—they should have been tossed over the wall to the jungle ages ago!" Kromodeor curled out an eye as he spoke, and complacently surveyed the writhing cylinder of sinuous, supple power that was his own body.

"Better avoid contact work with them if possible," cautioned Wixill. "You might not be able to unwrap, and to touch one of them is almost unthinkable. Speaking of wrapping, you know that they are putting on the finals of the contact work in the star this evening. Let's watch them."

They slid to the floor and wriggled away in perfect "step"—undulating along in such nice synchronism that their adjacent sides, only a few inches apart, formed two waving rigidly parallel lines. Deep in the lower part of the fortress they entered a large assembly room, provided with a raised platform in the center and having hundreds of short, upright posts in lieu of chairs; most of which were already taken by spectators. The two officers curled their tails comfortably around two of the vacant pillars, elevated their heads to a convenient level of sight and directed each an eye or two upon the stage. This was, of course, heptagonal. Its sides, like those of the mighty flying forts themselves, were not straight, but angled inward sufficiently to make the platform a seven-pointed star. The edge was outlined by a low rail, and bulwark and floor were padded with thick layers of a hard but smooth and yielding fabric.

IN this star-shaped ring two young Vorkuls were contending for the championship of the fleet in a contest that seemed to combine most of the features of wrestling, boxing, and bar-room brawling, with no holds barred. Four hands of each of the creatures held heavy leather billies, and could be used only in striking with those weapons, the remaining hands being left free to employ as the owner saw fit. Since the sport was not intended to be lethal, however, the eyes and other highly vulnerable parts were protected by metal masks, and the wing ribs were similarly guarded by leathern shields. The guiding fins, being comparatively small and extremely tough, required no protection.

"We're just in time," Kromodeor whistled. "The main bout is nicely on. See anyone from the flagship? I might stake a couple of korpels that Sintris will paint the symbol upon his wing."

"Most of their men seem to be across the star," Wixill replied, and both beings fell silent, absorbed in the struggle going on in the ring.

It was a contest well worth watching. Wing crashed against mighty wing and the lithe, hard bodies snapped and curled this way and that, almost faster than the eye could follow, in quest of advantageous holds. Above the shrieking wails of the crowd could be heard the smacks and thuds of the eight flying clubs as they struck against the leather shields or against tough and scaly hides. For minutes the conflict raged, with no advantage apparent. Now the fighters were flat upon the floor of the star, now dozens of feet in the air above it, as one or the other sought to gain a height from which to plunge downward upon his opponent; but both stayed upon or over the star—to leave its boundaries was to lose disgracefully.

Then, high in air, the visiting warrior thought that he saw an opening and grappled. Wings crashed in fierce blows, hands gripped and furiously wrenched. Two powerful bodies, tapering smoothly down to equally powerful tails, corkscrewed around each other viciously, winding up into something resembling tightly twisted lamp cord; and the two Vorkuls, each helpless, fell to the mat with a crash. Fast as was Zerexi, the gladiator from the flagship, Sintris was the merest trifle faster. Like the straightening of a twisted spring of tempered steel that long body uncoiled as they struck the floor, and up under those shielding wings—an infinitesimal fraction of a second slow in interposing—that lithe tail sped. Two lightning loops flashed around the neck of the visitor and tightened inexorably. Desperately the victim fought to break that terrible strangle hold, but every maneuver was countered as soon as it was begun. Beating wings, under whose frightful blows the very air quivered, were met and parried by wings equally capable. Hands and clubs were of no avail against that corded cable of sinew, and Sintris, his head retracted between his wings and his own hands reenforcing that impregnable covering over his head and neck, threw all his power into his tail—tightening, with terrific, rippling surges, that already throttling band about the throat of his opponent. Only one result was possible. Soon Zerexi lay quiet, and a violet beam of light flared from a torch at the ringside, bathing both contenders. At the flash the winner disengaged himself from the loser, and stood by until the latter had recovered the use of his paralyzed muscles. The two combatants then touched wing tips in salute and flew away together, over the heads of the crowd; plunging into a doorway and disappearing as the two officers uncoiled from their "seats" and wriggled out into the corridor.

"Fine piece of contact work," said Wixill, thoughtfully. "I'm glad that Sintris won, but I did not expect him to win so easily. Zerexi shouldn't have gone into a knot so early against such a fast man."

"Oh, I don't know," argued Kromodeor. "His big mistake was in that second body check. If he had blocked the sixth arm with his fifth, taken out the fourth and second with his third, and then gone in with...." and so, quite like two early experts after a good boxing match, the friends argued the fine points of the contest long after they had reached their quarters.

Day after day the vast duplex cone of Vorkulian fortresses sped toward the north pole of the great planet, with a high and constant velocity. Day after day the complex geometrical figure in space remained unchanged, no unit deviating measurably from its precise place in the formation. Over rapacious jungles, over geysers spouting hot water, over sullenly steaming rivers and seas, over boiling lakes of mud, and high over gigantic volcanoes, in uninterrupted eruptions of cataclysmic violence, the Vorkulian phalanx flew—straight north. The equatorial regions, considerably hotter than the poles, were traversed with practically no change in scenery—it was a world of steaming fog, of jungle, of hot water, of boiling, spurting mud, and of volcanoes. Not of such mild and sporadic volcanic outbreaks as we of green Terra know, but of gigantic primordial volcanoes, in terrifyingly continuous performances of frightful intensity. Due north the Vorkulian spearhead was hurled, before the rigorous geometrical alignment was altered.

"All captains, attention!" Finally, in a high latitude, the flagship sent out final instructions. "The hexans have detected us and our long range observers report that they are coming to meet us in force. We will now go into the whirl, and proceed with the maneuvers exactly as they have been planned. Whirl!"

At the command, each vessel began to pursue a tortuous spiral path. Each group of seven circled slowly about its own axis, as though each structure were attached rigidly to a radius rod, and at the same time spiraled around the line of advance in such fashion that the whole gigantic cone, wide open maw to the fore, seemed to be boring its way through the air.

"Lucky again!" Kromodeor, in the wardroom, turned to Wixill as the two prepared to take their respective watches. "It looks as though the first action would come while we're on duty. I've got just one favor to ask, if you have to economize on power, let Number One alone, will you?"

"No fear of that," Wixill hissed, with the Vorkulian equivalent of a chuckle. "We have abundance of power for all of your projector officers. But don't waste any of it, or I'll cut you down five ratings!"

"You're welcome. When I shine old Number One on any hexan work, one flash is all we'll take. See you at supper," and, leaving his superior at the door of the power room, Kromodeor wriggled away to his station upon the parallel horizontal bars before his panel.

Making sure that his tail coils were so firmly clamped that no possible lurch or shock could throw him out of position, he set an eye toward each of his sighting screens, even though he knew that it would be long before those comparatively short range instruments would show anything except friendly vessels. Then, ready for any emergency, he scanned his one "live" screen—the one upon which were being flashed the pictures and reports secured by the high-powered instruments of the observers.

WITH the terrific acceleration employed by the hexan spheres, it was not long until the leading squadron of fighting globes neared the Vorkulian war-cone. This advance guard was composed of the new, high-acceleration vessels. Their crews, with the innate blood-lust and savagery of their breed, had not even entertained the thought of accommodating their swifter pace to that of the main body of the fleet. These vast, slow-moving structures were no more to be feared than those similar ones whose visits they had been repulsing for twenty long Jovian years—by the time the slower spheres could arrive upon the scene there would be nothing left for them to do. Therefore, few in number as were the vessels of the vanguard, they rushed to the attack. In one blinding salvo they launched their supposedly irresistible planes of force—dazzling, scintillating planes under whose fierce power the studying, questing, scouting fortresses previously encountered had fled back southward; cut, beaten, and crippled. These spiraling monsters, however, did not pause or waver in their stolidly ordered motion. As the hexan planes of force flashed out, the dull green metal walls broke into a sparkling green radiance, against which the Titanic bolts spent themselves in vain. Then there leaped out from the weird brilliance of the walls of the fortresses great shafts of pale green luminescence—tractor ray after gigantic tractor ray, which seized upon the hexan spheres and drew them ruthlessly into the yawning open end of that gigantic cone.

Then, in each group of seven, similar great streamers of energy reached out from fortress to fortress, until each group was welded into one mighty unit by twenty-one such bands of force. The unit formed, a ray from each of its seven component structures seized upon a designated sphere, and under the combined power of those seven tractors, the luckless globe was literally snapped into the center of mass of the Vorkulian unit. There seven dully gleaming red pressor rays leaped upon it, backed by all the power of seven gigantic fortresses, held rigidly in formation by the unimaginable mass of the structures and by their twenty-one prodigious tractor beams. Under that awful impact, the screens and walls of the hexan spheres were exactly as effective as so many structures of the most tenuous vapor. The red glare of the vortex of those beams was lightened momentarily by a flash of brighter color, and through the foggy atmosphere there may have flamed briefly a drop or two of metal that was only liquefied. The red and green beams snapped out, the peculiar radiance died from the metal walls, and the gigantic duplex cone of the Vorkuls bored serenely northward—as little marked or affected by the episode as is a darting swift who, having snapped up a chance insect in full flight, darts on.

"Great Cat!" Far off in space, Brandon turned from his visiray screen and wiped his brow. "Czuv certainly chirped it, Perce, when he called those things flying fortresses. But who, what, why, and how? We didn't see any apparatus that looked capable of generating or handling those beams—and of course, when they got started, their screens cut us off at the pockets. Wish we could have made some sense out of their language—like to know a few of their ideas—find out whether we can't get on terms with them some way or other. Funny-looking wampuses, but they've got real brains—their think-tanks are very evidently full of bubbles. If they have it in mind to take us on next, old son, it'll be just ... too ... bad!"

"And then some," agreed Stevens. "They've got something—no fooling. It looks like the hexans are going to get theirs, good and plenty, pretty soon—and then what? I'd give my left lung and four front teeth for one long look at their controls in action."

"You and me both—it's funny, the way those green ray-screens stick to the walls, instead of being spherical, as you'd expect ... should think they'd *have* to radiate from a center, and so be spherical," Brandon cogitated. "However, we've got nothing corkscrewy enough to go through them, so we'll have to stand by. We'll stay inside whenever possible, look on from outside when we must, but all the time picking up whatever information we can. In the meantime, now that we've got our passengers, old Doctor Westfall prescribes something that he says is good for what ails us. Distance—lots of distance, straight out from the sun—and I wouldn't wonder if we'd better take his prescription."

The two Terrestrial observers relapsed into silence, staring into their visiray plates, searching throughout the enormous volume of one of those great fortresses in another attempt to solve the mystery of the generation and propagation of the incredible manifestations of energy which they had just witnessed. Scarcely had the search begun, however, when the visirays were again cut off sharply—the rapidly advancing main fleet of the hexans had arrived and the scintillant Vorkulian screens were again in place.

True to hexan nature, training and tradition, the fleet, hundreds strong, rushed savagely to the attack. Above, below, and around the far-flung cone the furious globes dashed, attacking every Vorkulian craft viciously with every resource at their command; with every weapon known to their diabolically destructive race. Planes of force stabbed and slashed, concentrated beams of annihilation flared fiercely through the reeking atmosphere, gigantic aerial bombs and torpedoes were hurled with full radio control against the unwelcome visitor—with no effect. Bound together in groups of seven by the mighty, pale-green bands of force, the Vorkulian units sailed calmly northward, spiraling along with not the slightest change in formation or velocity. The frightful planes and beams of immeasurable power simply spent themselves harmlessly against those sparklingly radiant green walls—seemingly as absorbent to energy as a sponge is to water, since the eye could not detect any change in the appearance of the screens, under even the fiercest blasts of the hexan projectors. Bombs, torpedoes, and all material projectiles were equally futile—they exploded harmlessly in the air far from their objectives, or disappeared at the touch of one of those dark, dull-red pressor rays. And swiftly, but calmly and methodically as at a Vorkulian practice drill, the heptagons were destroying the hexan fleet. Seven mighty green tractors would lash out, seize an attacking sphere, and snap it into the center of mass of the unit of seven. There would be a brief flash of dull red, a still briefer flare of incandescence, and the impalpable magnets would leap out to seize another of the doomed globes. It was only a matter of moments until not a hexan vessel remained; and the Vorkulian juggernaut spiraled onward, now at full acceleration, toward the hexan stronghold dimly visible far ahead of them—a vast city built around Jupiter's northern pole.

At the controls of his projector, Kromodeor spun a dial with a many-fingered, flexible hand and spoke.

"Wixill, I am being watched again—I can feel very plainly that strange intelligence watching everything I do. Have the tracers located him?"

"No, they haven't been able to synchronize with his wave yet. Either he is using a most minute pencil or, what is more probable, he is on a frequency which we do not ordinarily use. However, I agree with you that it is not a malignant intelligence. All of us have felt it, and none of us senses enmity. Therefore, it is not a hexan—it may be one of those strange creatures of the satellites, who are, of course, perfectly harmless."

"Harmless, but unpleasant," returned Kromodeor. "When we get back I'm going to find his beam myself and send a discharge along it that will end his spying upon me. I do not...."

AWAILING signal interrupted the conversation and every Vorkul in the vast fleet coiled even more tightly about his bars, for the real battle was about to begin. The city of the hexans lay before them, all her gigantic forces mustered to repel the first real invasion of her long and warlike history. Mile after mile it extended, an orderly labyrinth of spherical buildings arranged in vast interlocking series of concentric circles—a city of such size that only a small part of it was visible, even to the infra-red vision of the Vorkulians. Apparently the city was unprotected, having not even a wall. Outward from the low, rounded houses of the city's edge there reached a wide and verdant plain, which was separated from the jungle by a narrow moat of shimmering liquid—a liquid of such dire potency that across it, even those frightful growths could neither leap nor creep.

But as the Vorkulian phalanx approached—now shooting forward and upward with maximum acceleration, screaming bolts of energy flaming out for miles behind each heptagon as the full power of its generators was unleashed—it was made clear that the homeland of the hexans was far from unprotected. The verdant plain disappeared in a blast of radiance, revealing a transparent surface, through which could be seen masses of machinery filling level below level, deep into the ground as far as the eye could reach; and from the bright liquid of the girdling moat there shot vertically upward a coruscantly refulgent band of intense yellow luminescence. These were the hexan defences, heretofore invulnerable and invincible. Against them any ordinary warcraft, equipped with ordinary weapons of offense, would have been as pitifully impotent as a naked baby attacking a battleship. But now those defenses were being challenged by no ordinary craft; it had taken the mightiest intellects of Vorkulia two long lifetimes to evolve the awful engine of destruction which was hurling itself forward and upward with an already terrific and constantly increasing speed.

Onward and upward flashed the gigantic duplex cone, its entire whirling mass laced and latticed together—into one mammoth unit by green tractor beams and red pressors. These tension and compression members, of unheard-of power, made of the whole fleet of three hundred forty-three fortresses a single stupendous structure—a structure with all the strength and symmetry of a cantilever truss! Straight through that wall of yellow vibrations the vast truss drove, green walls flaming blue defiance as the absorbers overloaded; its doubly braced tip rearing upward, into and beyond the vertical as it shot through that searing yellow wall. Simultaneously from each heptagon there flamed downward a green shaft of radiance, so that the whole immense circle of the cone's mouth was one solid tractor beam, fastening upon and holding in an unbreakable grip mile upon mile of the hexan earthworks.

Practically irresistible force and supposedly immovable object! Every loose article in every heptagon had long since been stored in its individual shockproof compartment, and now every Vorkul coiled his entire body in fierce clasp about mighty horizontal bars: for the entire kinetic energy of the untold millions of tons of mass comprising the cone, at the terrific measure of its highest possible velocity, was to be hurled upon those unbreakable linkages of force which bound the trussed aggregation of Vorkulian fortresses to the deeply buried intrenchments of the hexans. The gigantic composite tractor beam snapped on and held. Inconceivably powerful as that beam was, it stretched a trifle under the incomprehensible momentum of those prodigious masses of metal, almost halted in their terrific flight. But the war-cone was not quite halted; the calculations of the Vorkulian scientists had been accurate. No possible artificial structure, and but few natural ones—in practice maneuvers entire mountains had been lifted and hurled for miles through the air—could have withstood the incredible violence of that lunging, twisting, upheaving impact. Lifted bodily by that impalpable hawser of force and cruelly wrenched and twisted by its enormous couple of angular momentum, the hexan works came up out of the ground as a waterpipe comes up in the teeth of a power shovel. The ground trembled and rocked and boulders, fragments of concrete masonry, and masses of metal flew in all directions as that city-encircling conduit of diabolical machinery was torn from its bed.

APORTION of that conduit fully thirty miles in length was in the air, a twisted, flaming inferno of wrecked generators, exploding ammunition, and broken and short-circuited high-tension leads before the hexans could themselves cut it and thus save the remainder of their fortifications. With resounding crashes, the structure parted at the weakened points, the furious upheaval stopped and, the tractor beams shut off, the shattered, smoking, erupting mass of wreckage fell in clashing, grinding ruin upon the city.

The enormous duplex cone of the Vorkuls did not attempt to repeat the maneuver, but divided into two single cones, one of which darted toward each point of rupture. There, upon the broken and unprotected ends of the hexan cordon, their points of attack lay: theirs the task to eat along that annular fortress, no matter what the opposition might bring to bear—to channel in its place a furrow of devastation until the two cones, their work complete, should meet at the opposite edge of the city. Then what was left of the cones would separate into individual heptagons, which would so systematically blast every hexan thing into nothingness as to make certain that never again would they resume their insensate attacks upon the Vorkuls. Having counted the cost and being grimly ready to pay it, the implacable attackers hurled themselves upon their objectives.

Here were no feeble spheres of space, commanding only the limited energies transmitted to their small receptors through the ether. Instead there were all the offensive and defensive weapons developed by hundreds of generations of warrior-scientists; wielding all the incalculable power capable of being produced by the massed generators of a mighty nation. But for the breach opened in the circle by the irresistible surprise attack, they would have been invulnerable, and, hampered as they were by the defenseless ends of what should have been an endless ring, the hexans took heavy toll.

The heptagons, massive and solidly braced as they were, and anchored by tractor rays as well, shuddered and trembled throughout their mighty frames under the impact of fiercely driven pressor beams. Sullenly radiant green wall-screens flared brighter and brighter as the Vorkulian absorbers and dissipators, mighty as they were, continued more and more to overload; for there were being directed against them beams from the entire remaining circumference of the stronghold. Every deadly frequency and emanation known to the fiendish hexan intellect, backed by the full power of the city, was poured out against the invaders in sizzling shrieking bars, bands, and planes of frenzied incandescence. Nor was vibratory destruction alone. Armor-piercing projectiles of enormous size and weight were hurled—diamond-hard, drill-headed projectiles which clung and bored upon impact. High-explosive shells, canisters of gas, and the frightful aerial bombs and radio-dirigible torpedoes of highly scientific war—all were thrown with lavish hand, as fast as the projectors could be served. But thrust for thrust, ray for ray, projectile for massive projectile, the Brobdingnagian creations of the Vorkuls gave back to the hexans.

The material lining of the ghastly moat was the only substance capable of resisting the action of its contents, and now, that lining destroyed by the uprooting of the fortress, that corrosive, brilliantly mobile liquid cascaded down in to the trough and added its hellish contribution to the furious scene. For whatever that devouring fluid touched flared into yellow flame, gave off clouds of lurid, strangling vapor, and disappeared. But through yellow haze, through blasting frequencies, through clouds of poisonous gas, through rain of metal and through storm of explosive the two cones ground implacably onward, their every offensive weapon centered upon the fast-receding exposed ends of the hexan fortress. Their bombs and torpedoes ripped and tore into the structure beneath the invulnerable shield and exploded, demolishing and hurling aside like straws, the walls, projectors, hexads and vast mountains of earth. Their terrible rays bored in, softening, fusing, volatilizing metal, short-circuiting connections, destroying life far ahead of the point of attack; and, drawn along by the relentlessly creeping composite tractor beam, there progressed around the circumference of the hexan city two veritable Saturnalia of destruction—uninterrupted, cataclysmic detonations of sound and sizzling, shrieking, multi-colored displays of pyrotechnic incandescence combining to form a spectacle of violence incredible.

But the heptagons could not absorb nor radiate indefinitely those torrents of energy, and soon one greenishly incandescent screen went down. Giant shells pierced the green metal walls, giant beams of force fused and consumed them. Faster and faster the huge heptagon became a shapeless, flowing mass, its metal dripping away in flaming gouts of brilliance; then it disappeared utterly in one terrific blast as some probing enemy ray reached a vital part. The cone did not pause nor waver. Many of its component units would go down, but it would go on—and on and on until every hexan trace had disappeared or until the last Vorkulian heptagon had been annihilated.

In one of the lowermost heptagons, one bearing the full brunt of the hexan armament, Kromodeor reared upright as his projector controls went dead beneath his hands. Finding his communicator screens likewise lifeless, he slipped to the floor and wriggled to the room of the Chief Power Officer, where he found Wixill idly fingering his controls.

"Are we out?" asked Kromodeor, tersely.

"All done," the Chief Power Officer calmly replied. "We have power left, but we cannot use it, as they have crushed our screens and are fusing our outer walls. Two out of seven chances, and we drew one of them. We are still working on the infra band, over across on the Second's board, but we won't last long...."

AS he spoke, the mighty fabric lurched under them, and only their quick and powerful tails, darting in lightning loops about the bars, saved them from being battered to death against the walls as the heptagon was hurled end over end by a stupendous force. With a splintering crash it came to rest upon the ground.

"I wonder how that happened? They should have rayed us out or exploded us," Kromodeor pondered. The Vorkuls, with their inhumanly powerful, sinuous bodies, were scarcely affected by the shock of that frightful fall.

"They must have had a whole battery of pressors on us when our greens went out—they threw us half-way across the city, almost into the gate we made first," Wixill replied, studying the situation of the vessel in the one small screen still in action. "We aren't hurt very badly—only a few holes that they are starting to weld already. When the absorber and dissipator crews get them cooled down enough so that we can use power again, we'll go back."

But they were not to resume their place in the attack. Through the holes in the still-glowing walls, hexan soldiery were leaping in steady streams, fighting with the utmost savagery of their bloodthirsty natures, urged on by the desperation born of the knowledge of imminent defeat and total destruction. Hand-weapons roared, flashed, and sparkled; heavy bars crashed and thudded against crunching bones; mighty bodies and tails whipped crushingly about six-limbed forms which wrenched and tore with monstrously powerful hands and claws. Fiercely and valiantly the Vorkuls fought, but they were outnumbered by hundreds and only one outcome was possible.

Kromodeor was one of the last to go down. Weapons long since exhausted, he unwrapped his deadly coils from about a dead hexan and darted toward a storeroom, only to be cut off by a horde of enemies. Throwing himself down a vertical shaft, he flew toward a tiny projector-locker, in the lowermost part of one of the great star's points, the hexans in hot pursuit. He wrenched the door open, and even while searing planes of force were riddling his body, he trained the frightful weapon he had sought. He pressed the contact, and bursts of intolerable flame swept the entire passage clear of life. Weakly he struggled to go out into the aisle, but his muscles refused to do the bidding of his will and he lay there, twitching feebly.

In the power room of the heptagon a hexan officer turned fiercely to another, who was offering advice.

"Vorkuls? Bah!" he snarled, viciously. "Our race is finished. Die we must, but we shall take with us the one enemy, who above all others needs destruction!" and he hurled the captured Vorkulian fortress into the air.

As the heptagon lurched upward, the massive door of a lower projector locker clanged shut and Kromodeor collapsed in a corner, his consciousness blotted out.

"WELL, that certainly tears it! That's a ... I...." Stevens' ready vocabulary failed him and he turned to Brandon, who was still staring narrow-eyed into the plate, watching the destruction of the hexan city.

"They've got something, all right—you've got to hand it to them," Brandon replied. "And we thought we knew something about forces and physical phenomena in general. Those birds have forgotten more than we ever will know. Just one of those things could take the whole I-P fleet, armed as we are now, any morning before breakfast, just for setting-up exercises. We've got to do something about it—but what?"

"It's okay—whatever you say. There may be an out somewhere, but I don't see it," and Stevens' gloomy tone matched his words.

Highly trained scientists both, they had been watching that which transcended all the science of the inner planets and knew themselves outclassed immeasurably.

"Only one thing to do, as I see it," Brandon cogitated. "That's to keep on going straight out, the way we're headed now. We'd better call a council of war, to dope out a line of action."

2 In order to avoid all unnecessary strain upon the memory of the reader, all titles, etc., have been given in the closest possible English equivalent, instead of in an attempted transliteration of the foreign word. This particular officer has no counterpart upon Tellurian vessels. He is the second in command of a Vorkulian fortress, his function being to supervise all expenditure of power.—E. E. S.

CHAPTER XII
The Citadel in Space

FOR the first time in many days Brandon and Westfall sat at dinner in the main dining room of the *Sirius*. They were enjoying greatly the unaccustomed pleasure of a leisurely, formal meal; but still their talk concerned the projection of pure forces instead of subjects more appropriate to the table; still their eyes paid more attention to diagrams drawn upon scraps of paper than to the diners about them.

"But I tell you, Quince, you're full of little red ants, clear to the neck!" Brandon snorted, as Westfall waved one of his arguments aside. "You must have had help to get that far off—no one man could possibly be as wrong as you are. Why, those fields absolutely will...."

"Hi, Quincy! Hi, Norman!" a merry voice interrupted. "Still fighting as usual, I see! What kind of knights are you, anyway, to rescue us poor damsels in distress, and then never even know that we're alive?" A tall, willowy brunette had seen the two physicists as she entered the saloon, and came over to their table, a hand outstretched to each in cordial greeting.

"Ho, Verna!" both men exclaimed, and came to their feet as they welcomed the smiling, graceful newcomer.

"Sit down here, Verna—we have hardly started," Westfall invited, and Brandon looked at the girl in assumed surprise as she seated herself in the proffered chair.

"Well, Verna, it's like this...." he began.

"That's enough!" she broke in. "That phrase always was your introduction to one of the world's greatest brainstorms. But I know that this is the first time you have had time even to eat like civilized beings, so I'll forgive you this once. Why all the registering of amazement, Norman?"

"I'm astonished that you aren't being monopolized by some husband or other. Surely the officers of the *Arcturus* weren't so dumb that they'd stand for your still being Verna *Pickering*, were they?"

"Not dumb, Norman, no. Far from it. But I'm still working for my M. R. S. degree, and I haven't succeeded in snaring it yet. You'd be surprised at how cagy those officers got after a few of them had been captured. But they are just like any other hunted game, I suppose—the antelopes that survive get pretty wild, you know," she concluded, plaintively.

"Well, that certainly is one tough break for a poor little girl," Brandon sympathized. "Quince, our little Nell, here, hasn't been done right by. I'm bashful and you're a woman-hater, but between us, some way, we've simply got to take steps."

"You might take longer steps than you think," Verna laughed, her regular, white teeth and vivid coloring emphasized by her olive skin and her startling hair, black as Brandon's own. "Perhaps I would like a scientist better than an I-P officer, anyway. The more I think of it, the surer I am that Nadia Newton had the right idea. I believe that I'll catch me a physicist, too—either of you would do quite nicely, I think," and she studied the two men carefully.

Westfall, the methodical and precise, had never been able to defend himself against Verna Pickering's badinage, but Brandon's ready tongue took up the challenge.

"Verna, if you really decided to get any living man he wouldn't stand a chance in the world," he declared. "If you've already made up your mind that I'm your meat, I'll come down like Davy Crockett's coon. But if either of us will do, that'll give us each a fifty-fifty chance to escape your toils. What say we play a game of freeze-out to decide it?"

"Fine, Norman! When shall we play?"

"Oh, between Wednesday and Thursday, any week you say," and the two fenced on, banteringly but skilfully, with Westfall an appreciative and unembarrassed listener.

Dinner over, Brandon and Westfall went back to the control room, where they found Stevens already seated at one of the master screens.

"All x, Perce?"

"All x. The observers report no registrations during the last two watches," and the three fell into discussion. Long they talked, studying every angle of the situation confronting them; until suddenly a speaker rattled furiously and an enormous, staring eye filled both master plates. Brandon's hand flashed to a switch, but the image disappeared even before he could establish the full-coverage ray screen.

"I'm on the upper band—take the lower!" he snapped, but Stevens' projector was already in action. Trained minds all, they knew that some intelligence had traced them, and all realized that it was of the utmost importance to know what and where that intelligence was. Stevens found the probing frequency in his range and they flashed their own beam along it, encountering finally one of the monstrous Vorkulian fortresses, far from Jupiter and almost directly between them and the planet! Its wall screens were in operation, and no frequency at their command could penetrate that neutralizing blanket of vibrations.

"What kind of an eye was that—ever see anything like it, Perce?" Brandon demanded.

"I don't think so, though of course we got only an awfully short flash of it. It didn't look like the periscopic eyes that those flying snakes had—looked more like a hexan eye, don't you think? Couldn't very well be hexan, though, in that kind of a ship."

"Don't think so, either. Maybe it's a purely mechanical affair that they use for observing. Anyway, old sons, I don't like the looks of things at all. Quince, you're the brains of this outfit—shift the massive old intellect into high and tell us what to do."

Westfall, staring into the eyepiece of the filar micrometer, finished measuring the apparent size of the heptagon before he turned toward Stevens and Brandon.

"It is hard to decide upon a course of action, since anything that we do may prove to be wrong," he said, slowly. "However, I do not see that this latest development can operate to change the plan we have already adopted; that of running away, straight out from the sun. We may have to increase our acceleration to the highest value the women and babies can stand. A series of observations of our pursuer will, of course, be necessary to decide that point. It would be useless to go to Titan, for they would be powerless to help us. We could not hold their mirror upon either the *Sirius* or their torpedoes against such forces as that fortress has at her command. Then, too, we might well be bringing down upon them an enemy who would destroy much of their world before he could be stopped. Both Uranus and Neptune are approximately upon our present course. Do the Titanians know anything of either of them, Steve?"

"Not a thing," the computer replied. "They can't get nearly as far as Uranus on their power beam—it's all they can do to make Jupiter. They seem to think, though, that one or more of the satellites of Uranus or Neptune may be inhabited by beings similar to themselves, only perhaps even more so. But considering the difference between what we found on the Jovian satellites and on Titan, I'd say that anything might be out there—on Uranus, Neptune, their satellites, or anywhere else."

"Cancel Uranus, and double that for Neptune," Brandon commanded. "Realize how far away they are?"

"That's right, too," agreed Stevens. "Before we got there, with any acceleration we can use now, this whole mess will be cleaned up, one way or the other."

WESTFALL completed the series of observations and calculated his results. Then, with a grave face, he went to consult the medical officers. The women, children, and the two Martian scientists were sent to the sick-bay and the acceleration was raised slowly to twenty meters per second per second, above which point the physicians declared they should not go unless it became absolutely necessary. Then the scientists met again—met without Alcantro and Fedanzo, who lay helpless upon narrow hospital bunks, unable even to lift their massive arms.

While Westfall made another series of precise measurements of the super-dreadnought of space so earnestly pursuing them, Brandon stumbled heavily about the room, hands jammed deep into pockets, eyes unseeing emitting clouds of smoke from his villainously reeking pipe. The Venetians, lacking Brandon's physical strength and by nature quieter of disposition, sat motionless; keen minds hard at work. Stevens sat at the calculating machine, absently setting up and knocking down weird and meaningless integrals, while he also concentrated upon the problem before them.

"They are still gaining, but comparatively slowly," Westfall finally reported. "They seem to be...."

"In that case we may be all x," Brandon interrupted, brandishing his pipe vigorously. "We know that they're on a beam—apparently we're the only ones hereabouts having cosmic power. If we can keep away from them until their beam attenuates, we can whittle 'em down to our size and then take them, no matter how much accumulator capacity they've got."

"But can we keep away from them that long?" asked Dol Kenor, pointedly; and his fellow Venerian also had a question to propound:

"Would it not be preferable to lead them in a wide circle, back to a rendezvous with the Space Fleet, which will probably be ready by the time of meeting?"

"I am afraid that that would be useless," Westfall frowned in thought. "Given power, that fortress could destroy the entire Fleet almost as easily as she could wipe out the *Sirius* alone."

"Kenor's right." Stevens spoke up from the calculator. "You're getting too far ahead of the situation. We aren't apt to keep ahead of them long enough to do much leading anywhere. The Titanians can hold a beam together from Saturn to Jupiter—why can't these snake-folks?"

"Several reasons," Brandon argued stubbornly. "First place, look at the mass of that thing, and remember that the heavier the beam the harder it is to hold it together. Second, there's no evidence that they wander around much in space. If their beams are designed principally for travel upon Jupiter, why should they have any extraordinary range? I say they can't hold that beam forever. We've got a good long lead, and in spite of their higher acceleration, I think we'll be able to keep out of range of their heavy stuff. If so, we'll trace a circle—only one a good deal bigger than the one Amonar suggested—and meet the fleet at a point where that enemy ship will be about out of power."

Thus for hours the scientists argued, agreeing upon nothing, while the Vorkulian fortress crept ever closer. At the end of three days of the mad flight, the pursuing space ship was in plain sight, covering hundreds of divisions of the micrometer screens. But now the size of the images was increasing with extreme slowness, and the scientists of the *Sirius* watched with strained attention the edges of those glowing green pictures. Finally, when the pictured edges were about to cease moving across the finely-ruled lines, Brandon cut down his own acceleration a trifle, and kept on decreasing it at such a rate that the heptagon still crept up, foot by foot.

"Hey what's the big idea?" Stevens demanded.

"Coax 'em along. If we run away from them they'll probably reverse power and go back home, won't they? Their beam is falling apart fast, but they're still getting so much stuff along it that we couldn't do a thing to stop them. If they think that we're losing power even faster than they are, though, they'll keep after us until their beam's so thin that they'll just be able to stop on it. Then they'll reverse or else go onto their accumulators—reverse, probably, since they'll be a long ways from home by that time. We'll reverse, too, and keep just out of range. Then, when we both have stopped and are about to start back, their beam will be at its minimum and we'll go to work on 'em—foot, horse, and marines. Nobody can run us as ragged as they've been doing and get away with it as long as I'm conscious and stand a chance in the world of hanging one onto their chins in retaliation. I've got a hunch. If it works, we can take those birds alone, and take 'em so they'll *stay* took. We might as well break up—this is going to be an ordinary job of piloting for a few days, I think. I'm going up and work with the Martians on that hunch. You fellows work out any ideas you want to. Watch 'em close, Mac. Keep kidding 'em along, but don't let them get close enough to puncture us."

EVERYTHING worked out practically as Brandon had foretold, and a few days later, their acceleration somewhat less than terrestrial gravity, he called another meeting in the control room. He came in grinning from ear to ear, accompanied by the two Martians, and seated himself at his complex power panel.

"Now watch the professor closely, gentlemen," he invited. "He is going to cut that beam."

"But you can't," protested Pyraz Amonar.

"I know you can't, ordinarily, when a beam is tight and solid. But that beam's as loose as ashes right now. I told you I had a hunch, and Alcantro and Fedanzo worked out the right answer for me. If I can cut it, Quince, and if their screens go down for a minute, shoot your visiray into them and see what you can see."

"All x. How much power are you going to draw?"

"Plenty—it figures a little better than four hundred thousand kilofranks. I'll draw it all from the accumulators, so as not to disturb you fellows on the cosmic intake. We don't care if we do run the batteries down some, but I don't want to hold that load on the bus-bars very long. However, if my hunch is right, I won't be on that beam five minutes before it's cut from Jupiter—and I'll bet you four dollars that you won't see the original crew in that fort when you get into it."

He set upper and lower bands of dirigible projectors to apply a powerful sidewise thrust, and the *Sirius* darted off her course. Flashing a minute pencil behind the huge heptagon, Brandon manipulated his tuning circuits until a brilliant spot in space showed him that he was approaching resonance with the heptagon's power beam. Micrometer dials were then engaged and the delicate tuning continued until the meters gave evidence that the two beams were precisely synchronized and exactly opposite in phase. Four plunger switches closed, that tiny pilot ray became an enormous rod of force, and as those two gigantic beams met in exact opposition and neutralized each other, a solid wall of blinding brilliance appeared in the empty ether behind the Vorkulian fortress. As that dazzling wall sprang into being, the sparkling green protection died from the walls of the heptagon.

"Go to it, Quince!" Brandon yelled, but the suggestion was entirely superfluous. Even before the wall-screen had died, Westfall's beam was trying to get through it, and when the visiray revealed the interior of the heptagon, the quiet and methodical physicist was shaken from his habitual calm.

"Why, they aren't the winged monsters at all—they're *hexans*!" he exclaimed.

"Sure they are." Brandon did not even turn his heavily-goggled eyes from the blazing blankness of his own screen. "That was my hunch. Those snakes went about things in a business-like fashion. They didn't strike me as being folks who would pull off such a wild stunt as trying to chase us clear out of the solar system, but a gang of hexans would do just that. Some of them must have captured that ship and, already having it in their cock-eyed brains that we were back of what happened on Callisto, they decided to bump us off if it was the last thing they ever did. That's what I'd do myself, if I were a hexan. Now I'll tell you what's happening back at the home power plant of that ship and what's going to happen next. I'm kicking up a horrible row out there with my interference, and a lot of instruments at the other end of that beam must be cutting up all kinds of didoes, right now. They'll check up on that ship with the expedition, by radio and what-not, and when they find out that it's clear out here—chop! Didn't get to see much, did you?"

"No, they must have switched over to their accumulators almost instantly."

"Yeah, but if they've got accumulator capacity enough to hold off our entire cosmic intake and get back to Jupiter besides, I'm a polyp! We're going to take that ship, fellows, and learn a lot of stuff we never dreamed of before. Ha! There goes his beam—pay me the four, Quince."

The dazzling wall of incandescence had blinked out without warning, and Brandon's beam bored on through space, unimpeded. He shut it off and turned to his fellows with a grin—a grin which disappeared instantly as a thought struck him and he leaped back to his board.

"Sound the high-acceleration warning quick, Perce!" he snapped, and drove in switch after switch.

"Cosmic intake's gone down to zero!" exclaimed MacDonald, as the *Sirius* leaped away.

"Had to cut it—they might shoot a jolt through that band. Just thought of something. Maybe unnecessary, but no harm done if ... it's necessary, all x—we're taking a sweet kissing right now. You see, even though we're at pretty long range, they've got some horrible projectors, and they were evidently mad enough to waste some power taking a good, solid flash at us—and if we hadn't been expecting it, that flash would have been a bountiful sufficiency, believe me—Great Cat! Look at that meter—and I've had to throw in number ten shunt! The outer screen is drawing five hundred and forty thousand!"

THEY stared at the meter in amazement. It was incredible, even after they had seen those heptagons in action, that at such extreme range any offensive beam could be driven with such unthinkable power—power requiring for its neutralization almost the full output of the prodigious batteries of accumulators carried by the *Sirius*! Yet for five, ten, fifteen, twenty minutes that beam drove furiously against their straining screens, and even Brandon's face grew tense and hard as that frightful attack continued. At the end of twenty-two minutes, however, the pointer of the meter snapped back to the pin and every man there breathed an explosive sigh of relief—the almost unbearable bombardment was over; the screen was drawing only its maintenance load.

"Wow!" Brandon shouted. "I thought for a minute they were going to hang to us until we cracked, even if it meant that they'd have to freeze to death out here themselves!"

"It would have meant that, too, don't you think?" asked Stevens.

"I imagine so—don't see how they could possibly have enough power left to get back to Jupiter if they shine that thing on us much longer. Of course, the more power they waste on us, the quicker we can take them; but I don't want much more of that beam, I'll tell the world—I just about had heart failure before they cut off!"

The massive heptagon was now drifting back toward Jupiter at constant velocity. The hexans were apparently hoarding jealously their remaining power, for their wall screens did not flash on at the touch of the visiray. Through unresisting metal the probing Terrestrial beams sped, and the scientists studied minutely every detail of the Vorkulian armament; while the regular observers began to make a detailed photographic survey of every room and compartment of the great fortress. Much of the instrumentation and machinery was familiar, but some of it was so strange that study was useless—days of personal inspection and experiment, perhaps complete dismantling, would be necessary to reveal the secrets hidden within those peculiar mechanisms.

"They're trying to save all the power they can—think I'll make them spend some more," Brandon remarked, and directed against the heptagon a heavy destructive beam. "We don't want them to get back to Jupiter until after we've boarded them and found out everything we want to know. Come here, Quince—what do you make of this?"

Both men stared at the heptagon, frankly puzzled; for the screens of the strange vessel did not radiate, nor did the material of the walls yield under the terrible force of the beam. The destructive ray simply struck that dull green surface and vanished—disappeared without a trace, as a tiny stream of water disappears into a partially-soaked sponge.

"Do you know what you are doing?" asked Westfall, after a few minutes' thought. "I believe that you are charging their accumulators at the rate of," he glanced at a meter, "exactly thirty-one thousand five hundred kilofranks."

"Great Cat!" Brandon's hand flashed to a switch and the beam expired. "But they can't just simply grab it and store it, Quince—it's impossible!"

"The word 'impossible' in that connection, coming from you, has a queer sound," Westfall said pointedly and Brandon actually blushed.

"That's right, too—we have got pretty much the same idea in our cosmic intake fields, but we didn't carry things half as far as they have done. Huh! They're flashing us again ... but those thin little beams don't mean anything. They're just trying to make us feed them some more, I guess. But we've got to hold them back some way—wonder if they can absorb a tractor field?"

The hexans had lashed out a few times with their lighter weapons, but, finding the *Sirius* unresponsive, had soon shut them off and were stolidly plunging along toward Jupiter. Brandon flung out a tractor rod and threw the mass of his cruiser upon it as it locked into those sullen green walls. But as soon as the enemy felt its drag, their screens flared white, and the massive Terrestrial space-ship quivered in every member as that terrific cable of force was snapped.

"They apparently cannot store up the energy of a tractor," commented Westfall, "but you will observe that they have no difficulty in radiating when they care to."

"Those two ideas didn't pan out so heavy. There's lots of things not tried yet, though. Our next best bet is to get around in front of him and push back. If they wiggle away from more than fifty percent of a pressor, they're really good."

The pilot maneuvered the *Sirius* into line, directly between Jupiter and the pentagon; and as the driving projectors went into action, Brandon drove a mighty pressor field along their axis, squarely into the center of mass of the Vorkulian fortress. For a moment it held solidly, then, as the screens of the enemy went into action, it rebounded and glanced off in sparkling, cascading torrents. But the hexans, with all their twisting and turning, could not present to that prodigious beam of force any angle sufficiently obtuse to rob it of half its power, and the driving projectors of the pentagon again burst into activity as the backward-pushing mass of the *Sirius* made itself felt. In a short time, however, the wall-screens were again cut off—apparently more power was required to drive them than they were able to deflect.

Although even the enormous tonnage of the Terrestrial cruiser was insignificant in comparison with the veritable mountain of metal to which she was opposed, so that the fiercest thrust of her driving projectors did not greatly affect the monster's progress; yet Brandon and his cohorts were well content.

"It's a long trip back to where they came from, and since they wanted to drift all the way, I think they'll be out of power before they get there,"

Brandon summed up the situation. "We aren't losing any power, either, since we are using only a part of our cosmic intake."

In a few hours the struggle had settled down to a routine matter—the *Sirius* being pushed backward steadily against the full drive of her every projector, contesting stubbornly every mile of space traversed. Assured that the regular pilots and lookouts were fully capable of handling the vessel, the scientists were about to resume their interrupted tasks when one of the photographers called them over to look at something he had discovered in one of the lowermost and smallest compartments of the heptagon. All crowded around the screens, and saw pictured there the winged, snake-like form of one of the original crew of the Vorkulian vessel!

"Dead?" Brandon asked.

"Not yet," replied the photographer. "He is twitching a little once in a while, but you see, he's pretty badly cut up."

"I see he is ... he must have a lot of vitality to have lasted this long—may be he'll live through it yet. Hold him on the plate, and get his exact measurements." He turned to the communicator. "Doctor von Steiffel? Can you come down to the control room a minute? We may want you to operate upon one of these South Jovians after a while."

"*Himmel! Es ... ist ... der....*" The great surgeon, bearded and massive, stared into the plate, and in his surprise started to speak in his native German. He paused, his long, powerful fingers tracing the likeness of the Vorkul upon the plate, then went on: "I would like very much to operate, but, not understanding our intentions, he would, of course, struggle. And when that body struggles—*schrecklichkeit*!" and he waved his arms in a pantomime of wholesale destruction.

"I thought of that—that's why I am talking to you now instead of when we get to him, two or three days from now. We'll give you his exact measurements, and a crew of mechanics will, under your direction, sink holes in the steel floor and install steel bands heavy enough to hold him rigid, from tailfins to wing-tips. We'll hold him there until we can make him understand that we're friends. It is of the utmost importance to save that creature's life if possible; because we do not want one of their fortresses launched against us—and in any event, it will not do us any harm to have a friend in the City of the South."

"Right. I will also have prepared some kind of a space-suit in which he can be brought from his vessel to ours," and the surgeon took the measurements and went to see that the "operating table" and suit were made ready for Kromodeor, the sorely wounded Vorkul.

IT was not long until the projectors of the heptagon went out and she lay inert in space, power completely exhausted. Knowing that the screens of the enemy would absorb any ordinary ray, the scientists had calculated the most condensed beam they could possibly project, a beam which, their figures showed, should be able to puncture those screens by sheer mass action—puncture them practically instantaneously, before the absorbers could react. To that end they had arranged their circuits to hurl seven hundred sixty-five thousand kilofranks—the entire power of their massed accumulators and their highest possible cosmic intake—in one tiny bar of superlative density, less than one meter in diameter! Everything ready, Brandon shot in prodigious switches that launched that bolt—a bolt so vehement, so inconceivably intense, that it seemed fairly to blast the very ether out of existence as it tore its way along its carefully predetermined line. The intention was to destroy all the control panels of the absorber screens; parts so vital that without them the great vessel would be helpless, and yet items which the Terrestrials could reconstruct quite readily from their photographs and drawings.

As that irresistible bolt touched the Vorkulian wall-screen, the spot of contact flared instantaneously through the spectrum and into the black beyond the violet as that screen overloaded locally. Fast as it responded and highly conductive though it was, it could not handle that frightfully concentrated load. In the same fleeting instant of time every molecule of substance in that beam's path flashed into tenuous vapor—no conceivable material could resist or impede that stabbing stiletto of energy—and the main control panel of the Vorkulian wall-screen system vanished. Time after time, as rapidly as he could sight his beam and operate his switches, Brandon drove his needle of annihilation through the fortress, destroying the secondary controls. Then, the walls unresisting, he cut in the vastly larger, but infinitely less powerful, I-P ray, and with it systematically riddled the immense heptagon. Out through the gaping holes in the outer walls rushed the dense atmosphere of Jupiter, and the hexans in their massed hundreds died.

The *Sirius* was brought up beside the heptagon, so that her main air-lock was against one of the yawning holes in the green metal wall of the enemy. There she was anchored by tractor beams, and the two hundred picked men of the I-P police, in full space equipment, prepared to board the gigantic fortress of the void. Brandon sat tense at his controls, ready to send his beam ahead of the troopers against any hexans that might survive in some as yet unpunctured compartment. General Crowninshield sat beside the physicist at an auxiliary board, phones at ears and four infra-red visiray plates ranged in front of him; ready through light or darkness to direct and oversee the attack, no matter where it might lead or how widely separated the platoons might become before the citadel was taken.

The space-line men—the engineers of weightless combat—led the van, protected by the projectors of their fellows. Theirs the task to set up ways of rope, along which the others could advance. Power drills bit savagely into metal, making holes to receive the expanding eyebolts; grappling hooks seized fast every protuberance and corner; points of little stress were supported by powerful suction cups; and at intervals were strung beam-fed lanterns, illuminating brilliantly the line of march. Through compartments and down corridors they went, bridging the many gaps in the metal through which Brandon's beams had blasted their way; guided by Crowninshield along the shortest feasible path toward the little projector room in which Kromodeor, the wounded Vorkul, lay. There were so many chambers and compartments in the heptagon that it had, of course, been impossible to puncture them all, and in some of the tight rooms were groups of hexans, anxious to do battle. But the general's eye led his men, and if such a room lay before them, Brandon's frightful beam entered it first—and where that beam entered, life departed.

But the hexans were really intelligent, as has been said. They had had time to prepare for what they knew awaited them, and they were rendered utterly desperate by the knowledge that, no matter what might happen, their course was run. Their power was gone, and even if the present enemy should be driven off, they would float idly in space until they died of cold; or, more probably, hurtling toward Jupiter as they were, they would plunge to certain death upon its surface as soon as they came within its powerful gravitational field. Therefore some fifty of the creatures, who had had space experience in their spherical vessels, had spent the preceding days in manufacturing space equipment. Let the weight-fiends plan upon detonating magazines of explosives, upon laying mines calculated to destroy the invaders, even the vessel itself and all within it. Let them plan upon any other such idle schemes, which were certain to be foreseen and guarded against by the space-hardened veterans who undoubtedly moaned that all-powerful and vengeful football of scarred gray metal. Space-fighters were they, and as space-fighters would they die; taking with them to their own inevitable death a full quota of the enemy.

THUS it came about that the head of the column of police had scarcely passed a certain door, when in the room behind it there began to assemble the half-hundred spacehounds of the hexans. When the vanguard had approached that room, Crowninshield had inspected it thoroughly with his infra-red beams. He had found it punctured and airless, devoid of life or of lethal devices, and had passed on. But now the space-suited warriors of the horde, guided in their hiding by their own visirays, were massing there. When the center of the I-P column reached that door, it burst open. There boiled out into the corridor, into the very midst of the police, fifty demoniacal hexans, fighting with Berserk fury, ruled by but one impulse—to kill.

Hand-weapons flashed viciously, tearing at steel armor and at bulging space-suits. Space-hooks bit and tore. Pikes and lances were driven with the full power of brawny arms. Here and there could be seen trooper and hexan, locked together in fierce embrace far from any hand-line—six limbs against four, all ten plied with abandon in mortal, hand-to-hand, foot-to-foot combat.

"Give way!" yelled Crowninshield into the ears of his men. "Epstein, back! LeFevre, advance! Get out of block ten—give us a chance to use a beam!"

As the police fell back out of the designated section of the corridor, Brandon's beam tore through it, filling it from floor to ceiling with a volume of intolerable energy. In that energy walls, doorway, and space-lines, as well as most of the hexans, vanished utterly. But the beam could not be used again. Every surviving enemy had hurled himself frantically into the thickest ranks of the police and the battle raged fiercer than ever. It did not last long. The ends of the column had already closed in. The police filled the corridor and overflowed into the yawning chasm cut by the annihilating ray. Outnumbered, surrounded upon all sides, above, and below by the Terrestrials, the hexans fought with mad desperation to the last man—and to the last man died. And even though in lieu of their own highly efficient space-armor they had fought in weak, crude, and hastily improvised space-suits, which were pitifully inferior to the ray resistant, heavy steel armor of the I-P forces, nevertheless the enormous strength and utter savagery of the hexans had taken toll; and when the advance was resumed, it was with extra lookouts scanning the entire neighborhood of the line of march.

Since the troops had entered the fortress as close to their goal as possible, it was not long until the leading platoon reached the door behind which Kromodcor lay. Tools and cylinders of air were brought up, and the engineers quickly fitted pressure bulkheads across the corridor. There was a screaming hiss from the valves, the atmosphere in that walled-off space became dense, and mechanics attacked with their power drills the door of the projector room. It opened, and four husky orderlies rapidly but gently encased the long body of the Vorkul in the space-suit built especially to receive it. As that monstrous form in its weirdly bulging envelope was guided through the air-locks into the *Sirius*, Crowninshield barked orders into his transmitter and the police reformed. They would now systematically scour the fortress, to wipe out any hexans that might still be in hiding; to discover and destroy any possible traps or infernal machines which the enemy might have planted for their undoing.

Assured that the real danger to the *Sirius* was over and that his presence was no longer necessary, Brandon turned his controls over to an assistant and went up to the Venerian rooms, where von Steiffel and his staff were to operate upon the Vorkul. There, in the dense, hot air, but little different now from the atmosphere of Jupiter, Kromodeor lay; bolted down to the solid steel of the floor by means of padded steel straps. So heavy were the bands that he could not possibly break even one of them; so closely were they spaced that he could scarcely have moved a muscle had he tried. But he did not try—so near death was he that his mighty muscles did not even quiver at the trenchant bite of the surgeon's tools. Von Steiffel and his aides, meticulously covered with sterile gowns, hoods, and gloves, worked in most rigidly aseptic style; deftly and rapidly closing the ghastly wounds inflicted by the weapons of the hexans.

"Hi, Brandon," the surgeon grunted as he straightened up, the work completed. "I did not use much antiseptic on him. Because of possible differences in blood chemistry and in ignorance of his native bacteria, I depended almost wholly upon asepsis and his natural resistance. It is a good thing that we did not have to use an anaesthetic. He is in bad shape, but if we can feed him successfully, he may pull through."

"Feed him? I never thought of that. What d'you suppose he eats?"

"I have an idea that it is something highly concentrated, from his anatomy. I shall try giving him sugar, milk chocolate, something of the kind. First I shall try maple syrup. Being a liquid, it is easily administered, and its penetrating odor also may be a help."

ACAN of the liquid was brought in and to the amazement of the Terrestrials, the long, delicate antennae of the Vorkul began to twitch as soon as the can was opened. Motioning hastily for silence, von Steiffel filled a bowl and placed it upon the floor beneath Kromodeor's grotesque nose. The twitching increased, until finally one dull, glazed eye brightened somewhat and curled slowly out upon its slender pedicle, toward the dish. His mouth opened sluggishly and a long, red tongue reached out, but as his perceptions quickened, he became conscious of the strangers near him.

The mouth snapped shut, the eye retracted, and heaving, rippling surges traversed that powerful body as he struggled madly against the unbreakable shackles of steel binding him to the floor.

"*Ach, kindlein*!" The surgeon bent anxiously over that grotesque but frightened head; soothing, polysyllabic German crooning from his bearded lips.

"Here, let's try this—I'm good on it," Stevens suggested, bringing up the Callistonian thought exchanger. All three men donned headsets, and sent wave after wave of friendly and soothing thoughts toward that frantic and terrified brain.

"He's got his brain shut up like a clam!" Brandon snorted. "Open up, guy—we aren't going to hurt you! We're the best friends you've got, if you only knew it!"

"Himmel, und he iss himself killing!" moaned von Steiffel.

"One more chance that might work," and Brandon stepped over to the communicator, demanding that Verna Pickering be brought at once. She came in as soon as the air-locks would permit, and the physicist welcomed her eagerly.

"This fellow's fighting so he's tearing himself to pieces. We can't make him receive a thought, and von Steiffel's afraid to use an anaesthetic. Now it's barely possible that he may understand hexan. I thought you wasted time learning any of it, but maybe you didn't—see if you can make him understand that we're friends."

The girl flinched and shrank back involuntarily, but forced herself to approach that awful head. Bending over, she repeated over and over one harsh, barking syllable. The effect of that word was magical. Instantly Kromodeor ceased struggling, an eye curled out, and that long, supple tongue flashed down and into the syrup. Not until the last sticky trace had been licked from the bowl did his attention wander from the food. Then the eye, sparkling brightly now, was raised toward the girl. Simultaneously four other eyes arose, one directed at each of the men and the other surveying his bonds and the room in which he was. Then the Vorkul spoke, but his whistling, hissing manner of speech so garbled the barking sounds of the hexan words he was attempting to utter, that Verna's slight knowledge of the language was of no use. She therefore put on one of the headsets, motioning the men to do the same, and approached Kromodeor with the other, repeating the hexan word of friendly import. This time the Vorkul's brain was not sealed against the visitors and thoughts began to flow.

"You've used those things a lot," Brandon turned to Stevens in a quick aside. "Can you hide your thoughts?"

"Sure—why?"

"All I can think of is that power system of theirs, and he'd know what we were going to do, sure. And I'd better be getting at it anyway. So you can wipe that off your mind with a clear conscience—the rest of us will get everything they've got there. Your job's to get everything you can out of this bird's brain. All x?"

"All x."

"Why, you didn't put yours on!" Verna exclaimed.

"No, I don't think I'll have time. If I get started talking to him now, I'd be here from now on, and I've got a lot of work to do. Steve can talk to him for me—see you later," and Brandon was gone.

He went directly to the Vorkulian fortress, bare now of hexan life and devoid of hexan snares and traps. There he and his fellows labored day after day learning every secret of every item of armament and equipment aboard the heptagon.

"Did you finish up today, Norm?" asked Stevens one evening. "Kromodeor's coming to life fast. He's able to wiggle around a little now, and is insisting that we take off the one chain we keep on him and let him use a plate, to call his people."

"All washed up. Guess I'll go in and talk to him—you all say he's such an egg. With this stuff off my mind I can hide it well enough. By the way, what does he eat?" And the two friends set out for the Venerian rooms.

"Anything that's sweet, apparently, with just enough milk to furnish a little protein. Won't eat meat or vegetables at all—von Steiffel says they haven't got much of a digestive tract, and I know that they haven't got any teeth. He's already eaten most all the syrup we had on board, all of the milk chocolate, and a lot of the sugar. But none of us can get any kind of a raise out of him at all—not even Nadia, when she fed him a whole box of chocolates."

"No, I mean what does he eat when he's home?"

"It seems to be a sort of syrup, made from the juices of jungle plants, which they drag in on automatic conveyors and process on automatic machinery. But he's a funny mutt—hard to get. Some of his thoughts are lucid enough, but others we can't make out at all—they are so foreign to all human nature that they simply do not register as thoughts at all. One funny thing, he isn't the least bit curious about anything. He doesn't want to examine anything, doesn't ask us any questions, and won't tell us anything about anything, so that all we know about him we found out purely by accident. For instance, they like games and sports, and seem to have families. They also have love, liking, and respect for others of their own race—but they seem to have no emotions whatever for outsiders. They're utterly inhuman—I can't describe it—you'll have to get it for yourself."

"Did you find out about the Callistonians who went to see them?"

"Negatively, yes. They never arrived. They probably couldn't see in the fog and must have missed the city. If they tried to land in that jungle, it was just too bad!"

"That would account for everything. So they're strictly neutral, eh? Well, I'll tell him 'hi,' anyway." Now in the sickroom, Brandon picked up the headset and sent out a wave of cheery greeting.

To his amazement, the mind of the Vorkul was utterly unresponsive to his thoughts. Not disdainful, not inimical; not appreciative, nor friendly—simply indifferent to a degree unknown and incomprehensible to any human mind. He sent Brandon only one message, which came clear and coldly emotionless.

"I do not want to talk to you. Tell the hairy doctor that I am now strong enough to be allowed to go to the communicator screen. That is all." The Vorkul's mind again became an oblivious maze of unintelligible thoughts. Not deliberately were Kromodeor's thoughts hidden; he was constitutionally unable to interest himself in the thoughts or things of any alien intelligence.

"Well, that for that." A puzzled, thoughtful look came over Brandon's face as he called von Steiffel. "A queer duck, if there ever was one. However, their ship will never bother us, that's one good thing; and I think we've got about everything of theirs that we want, anyway."

The surgeon, after a careful examination of his patient, unlocked the heavy collar with which he had been restraining the over-anxious Vorkul, and supported him lightly at the communicator panel. As surely as though he had used those controls for years Kromodeor shot the visiray beam out into space. One hand upon each of the several dials and one eye upon each meter, it was a matter only of seconds for him to get in touch with Vorkulia. To the Terrestrials the screen was a gray and foggy blank; but the manifest excitement shrieking and whistling from the speaker in response to Kromodeor's signals made it plain that his message was being received with enthusiasm.

"They are coming," the Vorkul thought, and lay back, exhausted.

"Just as well that they're comin' out here, at that," Brandon commented. "We couldn't begin to handle that structure anywhere near Jupiter—in fact, we wouldn't want to get very close ourselves, with passengers aboard."

Such was the power of the Vorkulian vessels that in less than twenty hours another heptagon slowed to a halt beside the *Sirius* and two of its crew were wafted aboard.

They were ushered into the Venerian room, where they talked briefly with their wounded fellow before they dressed him in a space-suit, which they filled with air to their own pressure. Then all three were lifted lightly into the air, and without a word or a sign were borne through the air-locks of the vessel, and into an opening in the wall of the rescuing heptagon. A green tractor beam reached out, seizing the derelict, and both structures darted away at such a pace that in a few minutes they had disappeared in the black depths of space.

"Well—that, as I may have remarked before, is indisputably and conclusively that." Brandon broke the surprised, almost stunned, silence that followed the unceremonious departure of the visitors. "I don't know whether to feel relieved at the knowledge that they won't bother us, or whether to get mad because they won't have anything to do with us."

He sent the "All x" signal to the pilot and the *Sirius,* once more at the acceleration of Terrestrial gravity, again bored on through space.

CHAPTER XIII
Spacehounds Triumphant

NOW that the hexan threat that had so long oppressed the humanity of the *Sirius* was lifted, that dull gray football of armor steel was filled with relief and rejoicing as the pilot laid his course for Europa. Lounges and saloons resounded with noise as police, passengers, and such of the crew as were at liberty made merry. The control room, in which were grouped the leaders of the expedition and the scientists, was orderly enough, but a noticeable undertone of gladness had replaced the tense air it had known so long.

"Hi, men!" Nadia Stevens and Verna Pickering, arms around each other's waists, entered the room and saluted the group gaily before they became a part of it.

"'Smatter, girls—tired of dancing already?" asked Brandon.

"Oh, no—we could dance from now on," Verna assured him. "But you see, Nadia hadn't seen that husband of hers for fifteen minutes, and was getting lonesome. Being afraid of all you men, she wanted me to come along for moral support. The real reason I came, though," and she narrowed her expressive eyes and lowered her voice mysteriously, "is that you two physicists are here. I want to study my chosen victims a little longer before I decide over which of you to cast the spell of my fatal charm."

"But you can't do that," he objected, vigorously. "Quince and I are going to settle that ourselves some day—by shooting dice, or maybe each other, or...." he broke off, listening to an animated conversation going on behind them.

"... just simply outrageous!" Nadia was exclaiming. "Here we saved his life, and I fed him a lot of my candy, and we went to all the trouble of bringing their ship back here almost to Jupiter for them, and then they simply dashed off without a word of thanks or anything! And he always acted as though he never wanted to see or hear of any of us again, ever! Why, they don't *think* straight—as Norman would say it, they're *full* of little red *ants*! Why, they aren't even *human*!"

"Sure not." Brandon turned to the flushed speaker. "They couldn't be, hardly, with their make-up. But is it absolutely necessary that all intelligent beings should possess such an emotion as gratitude? Such a being without it does seem funny to us, but I can't see that its lack necessarily implies anything particularly important. Keep still a minute," he went on, as Nadia tried to interrupt him, "and listen to some real wisdom. Quince, *you* tell 'em."

"They are, of course, very highly developed and extremely intelligent; but it should not be surprising that intelligence should manifest itself in ways quite baffling to us human beings, whose minds work so differently. They are, however ... well, peculiar."

"I *won't* keep still!" Nadia burst out, at the first opportunity. "I don't want to talk about those hideous things any more, anyway. Come on, Steve, let's go up and dance!"

Crowninshield turned to Verna, with the obvious intention of leading her away, but Brandon interposed.

"Sorry, Crown, but this lady is conducting a highly important psychological research, so your purely social claims will have to wait until after the scientific work is done."

"Why narrow the field of investigation?" laughed the girl. "I'd rather widen it, myself—I might prefer a general, even to a physicist!"

They went up to the main saloon and joined the mêlée there, and after one dance with Verna—all he could claim in that crowd of men—Crowninshield turned to Brandon.

"You two seem to know Miss Pickering extraordinarily well. Would I be stepping on your toes if I give her a play?"

"Clear ether as far as we're concerned." Brandon shrugged his shoulders. "She's been kicking around under foot ever since she was knee high to a duck—we gave her her first lessons on a slide rule."

"Don't be dumb, Norman. That woman's a knock-out—a riot—a regular tri-planet call-out!"

"Oh, she's all x, as far as that goes. She's a good little scout, too—not half as dumb as she acts—and she's one of the squarest little aces that ever waved a plume; but as for *playing* her—too much like our kid sister."

"Good—me for her!" and they made their way back down to the control room.

Stevens, after his one dance with Nadia, had already returned. Brandon and Crowninshield found him seated at the calculating machine, continuing a problem which already filled several pages of his notebook.

"'Smatter, Steve? So glad to see a calculator and some paper that you can't let them alone?"

"Not exactly—just had a thought a day or so ago. Been computing the orbit of the wreckage of the *Arcturus* around Jupiter. Think we should salvage it—the upper half, at least. It was left intact, you know."

"H ... m ... m. That would be nice, all right. Dope enough?"

"Got the direction solid, from my own observations; the velocity's a pretty rough approximation though. But after allowing for my probable error, it figures an ellipse of low eccentricity, between the orbits of Io and Europa. Its period is short—about two days."

"Isn't it wonderful to have a brain?" Brandon addressed the room at large. "The kid's clever. Nobody else would have thought of it, except maybe Westfall. Let's see your figures. Um ... m ... m. According to that, we're within an hour of it, right now." He turned to the pilot and sketched rapidly.

"Get on this line here, please, and decelerate, so that the stuff'll catch up with us, and pass the word to the lookouts. Stevens and I will take the bow plates.

"That's a good idea," he went on to Stevens, as they took their places at main and auxiliary ultra-banks. "Lot of plunder in that ship. Instruments, boats, and equipment worth millions, besides most of the junk of the passengers—clothes, trunks, trinkets, and what-not. You're there, bucko!"

"Thanks, Chief," ... and they fell silent, watching the instruments carefully, and from time to time making computations from the readings of the acceleration and flight meters.

"There she is!" An alarm bell had finally sounded, the ultra-lights had flared out into space, and upon both screens there shone out images of the closely clustered wreckage of the *Arcturus*. But both men were more interested just then in the mathematics of the recovery than in the vessel itself.

"Missed it eight minutes of time and eleven divisions on the scale," reported Stevens. "Not so good."

"Not so bad either—I've seen worse computation." Thus lightly was dismissed a mathematical feat which, a few years earlier, before the days of I-P computers, would have been deemed worthy of publication in "The Philosophical Magazine."

DIRECTOR NEWTON was called in, and it was decided that the many small fragments of the vessel were not worth saving; that its upper half was all that they should attempt to tow the enormous distance back to Tellus. The pace of the *Sirius* was adjusted to that of the floating masses, and tractor beams were clamped upon the undamaged portion of the derelict, and upon the two slices from the nose of the craft. A couple of the larger fragments of wreckage were also taken, to furnish metal for the repairs which would be necessary. Acceleration was brought slowly up to normal, and the battle-scarred cruiser of the void, with her heavy burden of inert metal, resumed her interrupted voyage toward Europa; the satellite upon which the passengers and crew of the ill-fated *Arcturus* had been so long immured. On she bored through the ether, detector screens full out and greenly scintillant Vorkulian wall-screens outlining her football shape in weird and ghastly light; unafraid now of any possible surviving space-craft of the hexans.

But if the hexans detected her, they made no sign. Perhaps their fleet had been destroyed utterly; perhaps it had been impressed upon even their fierce minds that those sparkling green screens were not to be molested with impunity! The satellite was reached without event and down into the crater landing shaft the two enormous masses of metal dropped.

Callisto's foremost citizens were on hand to welcome the Terrestrial rescuers, and revelry reigned supreme in that deeply buried Europan community. All humanity celebrated. The Callistonians rejoiced because they were now freed from the age-old oppression of the hexan hordes; because they could once more extend their civilization over the Jovian satellites and live again their normal lives upon the surface of those small worlds.

The Terrestrials were almost equally enthusiastic in the reunion that marked the end of the long imprisonment of the refugees.

As soon as the hull of the *Arcturus* had been warmed sufficiently to permit inspection, its original passengers were allowed to visit it briefly, to examine and to reclaim their belongings. Of course, some damage had been done by the cold of interplanetary space, but in general everything was as they had left it. Stevens and Nadia were among the first permitted aboard. They went first to the control room, where Stevens found his bag still lying behind Breckenridge's desk, where he had thrown it when he first boarded the vessel. Then they made their way up to Nadia's stateroom, which they found in meticulous order and spotless in its cleanliness—there is neither dust nor dirt in space. Nadia glanced about the formal little room and laughed up at her husband.

"Funny, isn't it, sweetheart, how little we know what to expect? Just think how surprised I would have been, when I left this room, if I had been told that I would have a husband before I got back to it!"

Breckenridge's first thought was for his precious triplex automatic chronometer, which he found, of course, "way off"—six and three-tenths seconds fast. Having corrected the timepiece from that of the *Sirius*, he immersed himself in the other delicate instruments of his department—and he was easy to find from that time on.

Overcrowded as the *Sirius* already was, it was decided that the original complement of the *Arcturus* should occupy their former quarters aboard her during the return trip. To this end, corps of mechanics set to work upon the salvaged hulk. Heavy metal work was no novelty to the Callistonian engineers and mechanics, and the *Sirius* also was well equipped with metal-working machines and men. Thus the prow was welded; armored, insulating air-breaks were built along the stern, which was the plane of hexan cleavage, electrical connections were restored; and lastly, a set of the great Vorkulian wall-screen generators, absorbers, and dissipators was installed, with sufficient accumulator capacity for their operation. Director Newton studied this installation in silence for some time, then went in search of Brandon.

"I hadn't considered the possibility of being attacked again between here and Tellus, but there's always the chance," he admitted. "If you think that there is any danger, we will crowd them all into the *Sirius*. It will not be at all comfortable, but it will be better than having any more of us killed."

"With that outfit they'll be as safe as we will," the scientist assured him. "They can stand as much grief as we can. We'll do the fighting for the whole outfit from here, and anything we meet will have to take us before they can touch them. So they had better ride it there, where they'll have passengers' accommodations and be comfortable. As to danger, I don't know what to expect. They may all be gone and they may not. We're going to expect trouble every meter of the way in, though, and be ready for it."

Everything ready and thoroughly tested, and stream of power flowing into the *Arcturus* from the cosmic receptors of her sister ship, the passengers and their new possessions were moved into their former quarters. There was a brief ceremony of farewell, the doors of the airlocks were closed, the careful check-out was gone through, and the driving projectors of the *Sirius* lifted both great vessels up the shaft, slowly and easily. And after them, as long as they could be seen, stared the thousands of Callistonians who thronged the great shaft's floor. Many of the spectators were not, strictly speaking, Callistonians at all. They were really Europans, born and reared in that hidden city which was to have been the last stronghold of Callisto's civilization. In that throng were hundreds who had never before seen the light of the sun nor any of the glories of the firmament, hundreds to whom that brief glimpse was a foretaste of the free and glorious life which was soon to be theirs.

Up and up mounted that powerful tug-boat of space, with her heavy barge, falling smoothly upward at normal acceleration. Below her first Europa, then mighty Jupiter, became moons growing smaller and smaller. In their stateroom Nadia's supple waist writhed in the curve of Stevens' arm as she turned and looked up at him with sparkling eyes.

"Well, big fellow, how does it feel to be out of a job? Or are you going over there every day on a tractor beam to work, as Norman suggested?"

"Not on your sweet young life!" he exclaimed. "Norm thought he was kidding somebody, but it registered zero. It gives me the pip to loaf around when there's a lot of work to do, but this is entirely different. Nothing's driving us now, and a fellow's entitled to at least one honeymoon during his life. And what a honeymoon this is going to be, little spacehound of my heart! Nothing to do but love you all the way from here to Tellus! Whoopee!"

"Oh, there's a couple of other things to do," she reminded him gaily. "You've got to smoke a lot of good cigarettes, I must eat a lot of Delray's chocolates, and we both really should catch up on eating fancy cooking.

Speaking of eating, isn't that the second call for dinner? It *is*!" and they went along the narrow hall toward the elevator. To these two the long journey was to seem all too short.

Long though the voyage was, it was uneventful. The occupants of the two vessels were in constant touch with each other by means of the communicators, and there was also much visiting back and forth in person. Stevens and Nadia came often to the *Sirius*, and were accompanied frequently by Verna Pickering, who claimed anew her ancient right of "kicking around under foot," wherever Brandon and Westfall might chance to be—and at such times General Crowninshield was practically certain to appear. And upon days when the beautiful brunette did not appear, the commandant generally found it necessary to inspect in person something in the *Arcturus*.

Day after day passed, and even the new and ultra-powerful detector screens of the *Sirius* remained unresponsive and cold. Day after day the plates before the doubled lookouts and observers remained blank. Power flowed smoothly and unfailingly into the cosmic receptors, and the products of conversion were discharged with equal smoothness and regularity from the forty-five gigantic driving projectors. The tractor beam held its heavy burden easily and the generators functioned perfectly. And finally a planet began to loom up in the stern lookout plates.

Verna, the irrepressible, was in the control room of the *Sirius*, quarreling adroitly with Brandon and deftly flirting with Crowninshield. Glancing into the control screen she saw the planet in its end block, then studied the instruments briefly.

"We're heading for *Mars*!" she declared with conviction. "I thought it looked that way yesterday, but supposed it must be only apparent—a trick of piloting or something about the orbit. I thought of course you were taking us back home—but you can't *possibly* get to Tellus on any such course as this!"

"Sure not," Brandon replied easily. "Certainly it's Mars. Isn't that where the *Arcturus* started out for? Whoever said we were going to Tellus? Of course, if any of the passengers want to go right back the IPC will undoubtedly furnish transportation *gratis*. But paste this in your hat, Verna, for future reference—when spacehounds start out to go anywhere they *go* there, even if they have to spend a year or so on minus time to do it!"

Closer and closer they approached the red planet, swinging around in a wide arc in order to make their course coincide exactly with the pilot ray of check station M14, which was now precisely in its scheduled location in space. At the chief pilot's desk in the control room of the *Arcturus*, Breckenridge checked in with the station, then calculated rapidly the instant of their touching the specially-built bumper platforms of spring steel, hemp, and fiber which awaited them upon the Martian dock of the Interplanetary Corporation. Within range of the terminal, he plugged into it, waited until the tiny light flashed its green message of attention, and reported.

"IPV *Arcturus*; Breckenridge, Chief Pilot; trip number forty-three twenty-nine. Checking in—four hundred forty-six days, fifteen hours, eleven minutes, thirty-eight and seven-tenths seconds minus!"